W. T. WELFORD

PROFESSOR OF PHYSICS, IMPERIAL COLLEGE, LONDON

Optics

Oxford University Press · 1976

Oxford University Press, Ely House, London W.1

OXFORD LONDON GLASGOW NEW YORK
TORONTO MELBOURNE WELLINGTON CAPE TOWN
IBADAN NAIROBI DAR ES SALAAM LUSAKA ADDIS ABABA
KUALA LUMPUR SINGAPORE JAKARTA HONG KONG TOKYO
DELHI BOMBAY CALCUTTA MADRAS KARACHI

Casebound ISBN 0 19 851830 7
Paperback ISBN 0 19 851831 5

© Oxford University Press 1976

Printed in Great Britain
by Billing & Sons Limited, Guildford and Worcester

Preface

Present-day physics courses are under increasing pressure, on
the one hand to keep up with developments in fundamental physics
and on the other to cover a broad range of topics appropriate to
the interests of students who may never become professional
physicists. Thus the time available for optics in the first or
second year of an undergraduate course, as for other branches of
physics, decreases, and this has influenced my choice of topics
in this book; I have been very selective and, as can be seen
from the contents list, I have chosen material which is either
basic to the development of the optics of the visible spectrum or
which has interesting links with other kinds of optics or other
branches of physics. Some may be concerned about what is *not* to
be found in this book, e.g. measurement of the speed of light,
group velocity, standing waves, the envelope function for diffraction gratings, refractometry, Fresnel diffraction, and phase-
change effects in interferometry. These omissions might have been
dictated anyway by the agreed size of the Oxford Physics Series
texts, but I do not plead this as an excuse. The book as it stands
is intended as a reasonable selection of topics to be presented
to undergraduates, perhaps in their first term at University and
certainly having to cope with many other new things at the same
time.

I have tried to stress physical arguments, and in order
to reduce the mathematical complexity I have introduced the concept
of a complex amplitude in the first chapter. I have also used
the formalism of Fourier-transform theory freely, since this
illuminates and simplifies every branch of physics in which waves
appear; this may seem rather extreme for an elementary text, but
since simple experiments with lasers are most easily discussed in

terms of Fourier transforms it seems almost certain that students will meet the transform in their laboratory work and will grasp the basic ideas even if they have not been presented with a systematic formulation. However, sections 5.5 and 6.6 contain some more difficult Fourier-transform material, which could be omitted in very early courses. The main definitions and theorems of Fourier-transform theory needed are given, without proofs, in the Appendix.

Some of the problems at the end of each chapter amplify the text by introducing simple extensions of the main discussion.

I should like to thank my colleagues Dr. M.E. Barnett and Dr. R.W. Smith for their help with this book, mostly given unknowingly; many of their ideas about the teaching of optics have gone into it. Also I am very grateful to Professor E.J. Burge, who read the first draft, gave very valuable criticism, and made many useful suggestions, and to Miss Lesley Harwood, who prepared the index; and I thank the staff of Oxford University Press for their help during publication.

The quotations from James Joyce's *Ulysses* are by kind permission of the Society of Authors, as the literary representative of the Estate of James Joyce, and of The Bodley Head, as publishers.

Imperial College,   W.T.W.
London, 1975

Contents

1. WAVES, RAYS, AND PARTICLES.   1

   The electromagnetic spectrum. Power and energy. The
   complex exponential notation and the complex amplitude.
   Sources and detectors. Monochromatic and polychromatic
   fields. Waves, particles, and rays. Problems.

2. GEOMETRICAL OPTICS   20

   The use of geometrical optics. Rays, wavefronts,
   reflection, and refraction. Optical images with a thin
   lens. Multi-element lenses. The Lagrange invariant and
   the power transmitted by an optical system. Non-
   paraxial optics. Problems.

3. PROPAGATION OF WAVES; INTERFERENCE AND DIFFRACTION   39

   Interference of two beams. Interference with extended
   and polychromatic light sources. Diffraction. Diffraction
   in the far field. Interference, diffraction, and the
   photon picture. Problems.

4. POLARIZATION   62

   Everyday aspects. Kinds of polarized light. Production
   of polarized light. Polarization and interference.
   Problems.

5. IMAGE-FORMING INSTRUMENTS   74

   Instrument design. Telescopes. The human eye. The
   microscope. Images of extended objects. Problems.

6. INTERFEROMETERS AND SPECTROSCOPES   91

   Young's experiment; spatial coherence. Michelson's
   interferometer; temporal coherence. Prisms and gratings
   as dispersing elements. Dispersion, resolution, and
   light-gathering power of prisms and gratings. Multiple-
   beam interference. Spectroscopy in general. Problems.

7. LASER LIGHT   119

   Coherent light speckle. Holography. Hologram interferometry.

Holographic diffraction gratings. Spatial filtering.
Problems.

APPENDIX: The Fourier transform and some of its   137
          properties

REFERENCES AND FURTHER READING   145

ANSWERS TO NUMERICAL PROBLEMS   146

INDEX   149

1. Waves, rays, and particles

But what I am anxious to arrive at is it is one thing to invent for instance those rays Röntgen did, or the telescope like Edison, though I believe it was before his time, Galileo was the man I mean. The same applies to the laws, for example, of a farreaching natural phenomenon such as electricity...

James Joyce, Ulysses

THE ELECTROMAGNETIC SPECTRUM

For many purposes optics can be regarded as the study of visible light, although in fact this light forms but a small part of a great range or spectrum of radiation. The most familiar part of this spectrum (apart from visible light) is probably the radio region (wireless waves). The complete spectrum of electromagnetic (e.m.) waves is described in Chapter 1 of *Radiation and quantum physics* (OPS 3) by D.J.E. Ingram. The waves are classified according to their wavelength λ or their frequency ν and these are related by

$$\lambda\nu = \text{velocity of the wave.} \qquad (1.1)$$

Electromagnetic waves of all frequencies have the same velocity in vacuum, approximately 3.10^8 m s^{-1}; this universal constant is denoted by c.

We shall begin by describing light and other parts of the e.m. spectrum as electromagnetic waves, but this is only one possible description; light (as all other regions of the spectrum) has many properties which are better discussed in terms of other representations (e.g. rays or particles), and we shall have to consider these also.

An e.m. wave can be represented as in Fig.1.1. The graph represents the strength of the electric field in the wave at a given instant and at different points along the direction z of travel. Fig.1.2 shows the same thing in a more picturesque way;

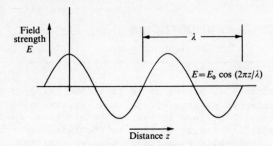

FIG.1.1. The electric field strength in an e.m. wave at a given instant as a function of the propagation distance z.

FIG.1.2. The amplitude of a wave. The closeness of the lines represents the field strength and broken lines indicate negative amplitudes.

the closeness of the lines indicates the relative strength of the electric field. Thus Figs 1.1 and 1.2 can be regarded as snapshots of the wave in space, taken at a certain instant of time. We could also look at a single point in space and consider the variation in time of the electric field at that point; we should then have a graph like Fig.1.3.

A more complete picture would be obtained by making the graph of Fig. 1.1 move along the z-axis at the velocity c of the wave. The field strength at any point as time passes would then vary as in Fig. 1.3. This travelling wave then has electric

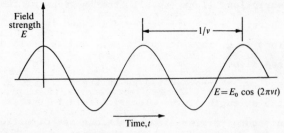

FIG. 1.3. The electric field strength in an e.m. wave at a given point as a function of time t

field strength E at any distance z and any time t given by

$$E = E_0\cos 2\pi(\nu t - z/\lambda). \qquad (1.2)$$

This is easily verified by keeping t or z constant and comparing with the expressions in Figs 1.1 and 1.3 respectively.

To complete the picture of an e.m. wave we ought to consider also the accompanying magnetic field. But here it is sufficient to note that the magnetic field has a similar sinusoidal variation and that in the simplest situations, where the wave is not trans-ferring energy to the medium through which it is travelling and where all parts of the wave are travelling in the same direction, the magnetic field varies in step or in phase with the electric field; both fields are at right-angles to the direction of travel of the wave.

Different sections of the e.m. spectrum are produced and detected in different ways, and the waves have a variety of interactions with matter, (see *Radiation and quantum physics* (OPS 3)). Although we shall be mainly concerned with visible light, it is easiest to consider first the properties of radio waves. This is because many of the properties we shall be interested in - those which produce interference and diffraction effects - can be demonstrated and explained for radio waves with fewer complications than for visible light.

POWER AND ENERGY

An essential property of all waves is that they transfer energy (from a source to a detector) without transferring the medium in which the waves occur. Indeed it is doubtful whether there can be said to be a "medium" for e.m. waves. Thus the rate of energy flow or the power in a wave is of interest. It follows from the detailed study of e.m. waves that for a wave like that in Figs 1.1 - 1.3 the power density (i.e. power per unit area across the wave transmitted in the direction of propagation) is proportional to the square of the electric field strength. We shall take this result as our starting point for a discussion of energy flow; it is treated in detail in texts on electromagnetism (e.g. *Electromagnetism* (OPS 1) by F.N.H. Robinson), where derivations and conditions of applicability can be found. Thus from (1.2) the power density is proportional to

$$E_0^2 \{1 + \cos 4\pi(\nu t - z/\lambda)\}. \tag{1.3}$$

Clearly the cosine term causes a periodic fluctuation in energy flow across a certain plane, say $z = 0$. The oscillating electric field induces an alternating voltage in a conductor (antenna), and this constitutes detection of the e.m. wave.

One of the major differences between e.m. waves at radio and at optical (and higher) frequencies is that we have no detectors which can respond fast enough to demonstrate optical frequencies directly. In fact the fastest detectors of light will respond only to frequencies of the order of $10^9 - 10^{10}$ Hz, some 5 orders of magnitude too low. Thus any detector of e.m. radiation in the optical range responds only to the average power over many cycles of the waves. This *time-averaged power* is thus (from (1.3)) proportional simply to E_0^2.

THE COMPLEX EXPONENTIAL NOTATION AND THE COMPLEX AMPLITUDE

Another basic property of e.m. waves is that if two or more wave systems cross in a certain region of space, the electric and magnetic field strengths in this region are found simply by adding as vectors the fields from the individual wave systems. Thus we find the effect of overlapping waves by adding their field strengths or by linear combination. This simple result is not true for very large field strengths, and the topic of *nonlinear optics* has developed in the last decade now that such field strengths are available at optical frequencies. However, in this book we, shall assume linear combination or *superposition*.

Both interference and diffraction phenomena can be explained in terms of superposition of waves, and in this section we shall discuss the mathematical symbolism for this.

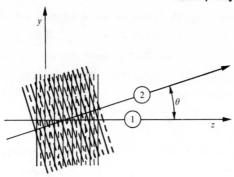

FIG. 1.4. The superposition of two e.m. waves travelling in .n
directions at an angle θ to each other.

Suppose we have two e.m. waves of the kind described on p.1,
travelling at an angle θ to each other, as in Fig. 1.4. Let the
two waves have the same frequency (and therefore the same wavelength) and the same maximum field strength E_0. If we use axes
as in the figure we can write the two waves as

$$E_0 \cos 2\pi (\nu t - z/\lambda),$$
$$E_0 \cos 2\pi \{\varepsilon + \nu t - (z\cos\theta + y\sin\theta)/\lambda\}. \qquad (1.4)$$

In the expression for the second wave the constant ε, known as
a phase-shift term, allows for the possibility that the two
waves are not in step at the origin of the coordinate system,
and the expression $z\cos\theta + y\sin\theta$ ensures that the lines of
constant electric field, or wavefronts, are at an angle θ to the
y-axis. To fix our ideas we can regard each of the parallel
lines in the figure as representing maximum field at a certain
instant of time, but this is not essential. In order to find
the *interference field*, as it is called, in the region where the
waves cross we have to add the two expressions (1.4). If we are
dealing with optical frequencies we can only observe the time-
averaged power density, which is, of course, what we ordinarily
know as the light intensity, and so we have to square the sum
of the two expressions in (1.4) and find the time-average.
There is no fundamental difficulty in doing this, but the
manipulation of the trigonometrical expressions is very involved,

particularly if we want to consider more than two waves and if they all have different field strengths. This has led to the introduction of the complex exponential notation and the use of the *complex amplitude* to describe waves, as follows.

First we replace an expression such as that in (1.2) by

$$E = E_0 \exp 2\pi i (\nu t - z/\lambda),$$

where i is, of course, $\sqrt{-1}$. We shall add complex expressions of this kind in superposing waves, but with the understanding that we are actually concerned only with the real parts. Since real and imaginary always remain separate in linear operations, this is valid. The above expression represents a wave with plane wavefronts travelling in the z-direction. We can now represent a similar plane wave, travelling in an arbitrary direction specified by a unit length vector $\underline{a}$, by the expression

$$E = E_0 \exp 2\pi i (\nu t - \underline{a} \cdot \underline{r}/\lambda),$$

where $\underline{r} = (x, y, x)$ is the vector from the origin to an arbitrary point in space. We can check that this agrees with the second of (1.4) by expanding the scalar product and remembering that the components of a unit vector are direction cosines.

Next we put $2\pi\nu = \omega$, the angular frequency, and we put $2\pi\underline{a}/\lambda = \underline{k}$. $\underline{k}$ is called the wave-vector, and we shall also use the scalar $|\underline{k}| = 2\pi/\lambda$, which we denote by k and call the wavenumber. Thus our expression for a plane wave is

$$E(t, \underline{r}) = E_0 \exp i (\varepsilon + \omega t - \underline{k} \cdot \underline{r}). \tag{1.5}$$

We have now indicated explicitly that the field strength E is a function of the position $\underline{r}$ and the time t, and we put in an arbitrary phase shift ε. We get the effect of superposing n waves of this kind by adding the appropriate terms,

$$\sum_n E_n \exp i (\varepsilon_n + \omega t - \underline{k}_n \cdot \underline{r}),$$

or, taking out the common factor $\exp i\omega t$, since we have supposed all the waves to have the same frequency,

$$\exp i\omega t \sum_n E_n \exp i (\varepsilon_n - \underline{k}_n \cdot \underline{r}).$$

We can write the summation, which is independent of the time, as $R + iI$, where R and I are two real functions of the position vector $\underline{r}$. From p.4 the intensity in the wave-field is the time-average of the square of the real part of

$$(R + iI) \exp i\omega t,$$

i.e, the time-average of

$$(R\cos\omega t - I\sin\omega t)^2.$$

It is easily verified that this time-average is simply $\frac{1}{2}(R^2 + I^2)$. The factor $\frac{1}{2}$ is usually dropped.

In this calculation the time-dependence of the waves appeared as a common factor $\exp i\omega t$ to all terms, which vanished in the final time-averaging; and the final intensity $R^2 + I^2$ is simply the squared modulus of the summed complex expressions.

Thus we have the rule that, to find the intensity due to several superposed plane waves of the same frequency, we add terms of the type $E_n \exp i(\varepsilon_n - \underline{k}_n \cdot \underline{r})$ for the individual waves and take the squared modulus at the end to find the intensity. An expression of the type

$$E\exp i(\varepsilon - \underline{k}\cdot\underline{r}),$$

in which the time-dependent part is omitted, is called a *complex amplitude*. These quantities can also be used for superposing other than plane waves (i.e. convergent or divergent waves), and for calculations with all forms of wave motion, not only e.m. waves. It is only necessary that the waves all have the same frequency. As a trivial example, the complex amplitude of the wave in eqn (1.2) is

$$E_0 \exp(-2\pi i z/\lambda),$$

and the intensity is therefore immediately E_0^2. If we now apply the procedure to the two waves of eqn (1.4) we easily find, for the intensity in the plane $z = 0$, the expression

$$2E_0^2 (1 + \cos\{(2\pi/\lambda)y\sin\theta\}).$$

This is a typical two-beam interference expression; we shall examine it more closely in Chapters 3 and 6.

As we noted earlier, the intensity, which has dimensions
of power per unit area, is strictly proportional to E_0^2, i.e.
in our present terms it is proportional to the squared modulus
of the complex amplitude. The proportionality constant is
important both for its dimensionality and for its numerical
magnitude in connection with radio wave and microwave theory,
but it is not important in the optical problems that we shall
encounter. Thus for many purposes we can ignore the
electromagnetic nature of light and discuss its properties in
terms of a complex amplitude of some undefined quality or medium.
Often we need not even specify whether the wave motion is
transverse (e.m. waves or surface waves on water) or longi-
tudinal (sound waves in air). This apparently abstract approach
has advantages: parallels with other kinds of wave can be drawn,
and we shall find it easier to come to terms with the fact that
even the electromagnetic theory is not adequate to explain all
optical phenomena.

It is found that all kinds of waves have to be characterized
by two different quantities. These are of widely differing
physical natures, depending on the kind of wave, but in all
cases there is an *amplitude*, which varies in time and space and
gives interference effects, and an *intensity*, which represents
the rate of energy transport. With suitable interpretations
the complex amplitude and its squared modulus, the intensity,
can be used in all cases. All interference experiments and
many diffraction experiments can be completely explained in
these terms.

SOURCES AND DETECTORS

The production and detection of different parts of the
e.m. spectrum are described in *Radiation and quantum physics*
(OPS 3). Many of the effects and techniques which we usually
call 'optical' apply mainly to the infrared, visible,

and ultraviolet regions. In these regions there are three
main kinds of source:

(1) thermal sources which produce a continuous spectrum,
     e.g. solid hot bodies, such as filament lamps, and
     hot gases under high pressure, as in an electric
     discharge through xenon (e.g. a flash tube);

(2) thermal sources giving line spectra, e.g. mercury
     vapour or neon discharge tubes under low pressure;

(3) lasers.

We can describe the production and detection of radio waves
quite well in terms of the classical theory of electromagnetism,
i.e. without invoking the existence of electrons or using quantum
theory. However, in the optical region of the spectrum we have
to introduce quantum concepts in order to explain light
production and detection, although effects concerned with
propagation alone (e.g. interference and diffraction) can be
described in terms of a simple wave theory, usually involving
only the use of the complex amplitude.

The quantum theory of light emission and absorption is
explained in *Radiation and quantum physics* (OPS 3). Here we need
only note that electromagnetic radiation is emitted or absorbed
in finite quanta of energy called *photons*. The amount of energy
in a photon depends on the frequency of the radiation and is
given by

$$E = h\nu = hc/\lambda \qquad (1.6)$$

where h, the Planck constant, is 6.626×10^{-34} J s. The energy
per photon is sometimes given in electronvolts (eV); 1 eV is
1.602×10^{-19} J. The emission or absorption of a photon
corresponds to a change in the energy of an atom, molecule,
or other system. In the infrared these transitions are between
rotational or vibrational states of molecules; in the visible
and near ultraviolet they correspond to changes in the energy
levels of electrons in the outer orbits of an atom; and in the

far ultraviolet and X-ray regions the inner electrons are
involved. These are progressively greater changes in energy
of a molecule or atom, and so they produce more energetic
photons. This affects the mode of detection. A far ultraviolet photon has enough energy to ionize a gas atom or
molecule, and it can be detected by an ionization chamber;
alternatively it can cause photoelectric emission of electrons
from almost any conductor, and so it can be detected by
photodiodes. Visible-light photons can produce photoelectrons
from particular surfaces (photocathodes), so that they can also
be detected photoelectrically; they can produce a latent image
in a photographic emulsion (a complex process which is only
understood in broad outline), and they can produce photochemical
reactions, one of which is the starting point of the process of
vision.

We call all the above processes *quantum detection processes*,
because they involve a change in state of an individual atom or
molecule by a single quantum. As we move up the wavelength
scale into the infrared, fewer quantum detectors are available
and the radiation is detected by its general heating effect when
it is absorbed, e.g. by a thermocouple or by a resistance
thermometer (*thermal detection processes*). The main reason
for this difference is that, in the infrared, the amount of
energy in a photon is so small as to be comparable to the average
random energy of thermal motion of the atoms or molecules in a
detector. This thermal energy is of order kT, where k is the
Boltzmann constant (1.281×10^{-23} J K^{-1}) and T is the absolute
temperature. Thus at ordinary temperatures the thermal energy
is about 1/40 eV, and we should not expect to be able to make
detectors depending on a quantum effect for photons of energy
less than or comparable to this value. We can improve the
quantum detection process by cooling the detector, and there are
a few devices in which the limitation is circumvented by

special procedures, but, broadly speaking, the limited photon
energy in this region of the spectrum is the reason for the
distinction between quantum and thermal detectors.

Detectors vary in their time-response to e.m. radiation.
Photocells and photomultipliers respond to changes taking place
as rapidly as 10^{-9} s; the human eye can only see changes slower
than about 0.05 s, an effect known to us as persistence of
vision; and photographic detectors can add or integrate light
flux for perhaps several hours, until a certain saturation of
exposure has been reached. We can also classify detectors
according to whether they record images or total flux. Photomultipliers, ionization chambers for X-rays, and thermocouples
for the infrared are *total flux detectors*; but the eye, the
photographic plate, and television camera tubes record detailed
images, i.e. they are *flux density detectors*.

It is often important to know whether a detector is linear
in response. The output of a detector can take many different
forms, such as a voltage, a current, blackening of a photographic emulsion, or a visual impression in the brain. If we
refer to any of these as the 'signal' the detector is said to
be linear in response if the signal is directly proportional
to the flux falling on the detector. Thus the signal
corresponding to the sum of two fluxes is the sum of the
individual signals. Photocells and photomultipliers are linear
over many orders of magnitude if they are used with suitable
circuits, but photographic detectors are usually not. The eye
is very nonlinear; the sensation of brightness is roughly
proportional to the logarithm of the light intensity.

A final interesting property of detectors is their spectral
range of sensitivity, or working range. This is roughly
indicated in e.m. spectrum charts (see *Radiation and quantum
physics* - OPS 3), but more information can be provided by a
graph. Such a graph may take several forms: (1) we can plot

the output signal per unit wavelength interval which the detector would give if used with a fictitious source giving the same energy flux per unit wavelength interval all over the spectrum; (2) we could make a similar plot but per unit frequency interval for detector and source; (3) for a quantum detector we can plot the reciprocal of the average number of photons required to produce one photoelectron as a function of wavelength or frequency, this being called the *quantum efficiency*. Fig. 1.5 shows spectral-sensitivity curves for the eye and for

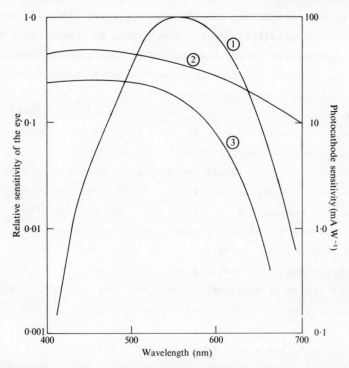

FIG. 1.5. Spectral sensitivity of the normal eye (curve 1), of an antimony-caesium photocathode (curve 2) and of a sodium-potassium-antimony-caesium photocathode (curve 3). The ordinate scale for the eye, on the left, is in arbitrary units, scaled to unity at the maximum sensitivity. The scale for the photo-cathodes, on the right, is in milliamperes of photocurrent per watt of incident light power.

some photocells. The difference in response over even the
narrow range of the visible spectrum complicates the comparison
of responses of different kinds of detectors. For example,
it can be seen that if we had beams with equal powers in watts
(W), but of violet and green light, an antimony-caesium
photocathode would suggest that the violet was the more intense,
but an eye would indicate the reverse. This has led to the use of
a special system of visual photometric units applicable to the
eye in which the *lumen* is the unit of flux (see Welford 1962);
a lumen is the equivalent of about $1 \cdot 467 \times 10^{-3}$ W of green
light or about $1 \cdot 2$W of violet light of wavelength $0 \cdot 410 \mu m$. In
this book we shall not use this visual system of units, since it
applies specifically to one detector; light flux will be measured
either in watts or in photons per second, with the wavelength
or frequency specified.

MONOCHROMATIC AND POLYCHROMATIC FIELDS

We must now consider an essential difference between radio
waves and light waves. On pp.2 and 5 we represented a radio
wave as a sinusoidal variation of electric field (see e.g.
eqn (1.2)), so that if a detector with sufficiently rapid response
were stationed at a fixed point in the field the output signal
would be a strictly sinusoidal or simple harmonic function of
time. This is very accurately true for unmodulated radio waves,
i.e. the simple carrier wave for radio or television. However,
it is not generally the case for visible light. If we could
examine the light vibrations carefully over a sufficient number
of cycles of the vibrations (and there are indirect ways of
doing this), we should find that, although the vibration seems
to be simple harmonic for short lengths of time, when examined
over longer periods the amplitude varies irregularly and the
maxima and minima do not recur at exactly equal time-intervals.
Fig. 1.6 suggests this effect of random variation of phase and
amplitude in a beam of light.

FIG. 1.6.   An almost monochromatic wave-train

The cause of the randomness in a light beam is easily seen.
A radio wave is generated by an alternating current (i.e. a
stream of electrons) in a conductor; and it is as if all the
electrons are in step all the time, to a very close approximation. However, a beam of light is the summation of a large
number of elementary waves (i.e. photons) emitted by atoms or
molecules, and in general there is no fixed relationship between
the times at which the different photons are emitted. Thus the
instantaneous amplitude in the beam of light is obtained as the
resultant of many independent waves of random phase but the
same frequency.

Under certain conditions we can still regard the light
beam as simple harmonic, and then it becomes fairly easy to
discuss interference and diffraction. It is for this reason
that we said earlier that it is simplest to speak in terms of
radio waves first. The two kinds of wave motion we have
described may be called *monochromatic* and *polychromatic*.

The actual lengths of time for which light beams can be
regarded as simple harmonic vary greatly. At one extreme we
have 'white light' from, for example, a tungsten filament lamp.
This light contains a continuous range of wavelengths - as was
shown by Isaac Newton, using a prism - and since the wavelengths
have random phases we would not expect the simple harmonic
property to persist for more than a few periods, i.e. about
10^{-14} s. In a 'monochromatic' beam, e.g. one of the spectrum
lines from a mercury discharge lamp, for several reasons the
photons have a (small) range of frequencies, and since they are

emitted with random phases there is again a finite time
involved. Depending on the temperature and pressure of the
discharge the time for which the wave persists as substantially
harmonic may be some thousands or tens of thousands of periods,
i.e. 10^{-11} - 10^{-10} s. By means of the velocity of light
(3×10^{8} m s^{-1}) we can express this in terms of the *length* of
the wave train which is substantially simple harmonic in form,
i.e. a few millimetres to a few tens of millimetres. These
times and distances can only be stated approximately, since the
wave-trains do not change abruptly but gradually, as in Fig. 1.6.
They are determined by experiments with an interferometer
(see p.97).

Laser light contains even longer stretches of simple harmonic
waves. This is because the mode of action of the laser constrains
the light-emitting atoms to emit photons which are precisely
in phase with each other, instead of having random phase
relationships, as in an ordinary light source. There are still
some random fluctuations, but in carefully stabilized lasers the
light may be truly simple harmonic for times as long as 10^{-6} s,
so that the wave-trains are about 1000 m long.

The time for which the wave-train remains simple harmonic
is called the *coherence time* and the corresponding distance, i.e.
this time multiplied by the velocity, is the *coherence length*.
These quantities determine the possibility of getting certain
interference effects. If we have two radio beams from two
aerials powered from the same radio-frequency transmitter, there
is always the same phase relationship between the electric
fields at a certain point. Thus in Fig.1.7 if A and B are the
two transmitters, and if P_1 is 4 wavelengths from A and 5
wavelengths from B, there will be a powerful signal at P_1 in a
receiver, since the two waves add in phase. However, at P_2,
which is 4.5 wavelengths from A and 5 wavelengths from B, there
will be nearly zero signal. It is easy to map regions of

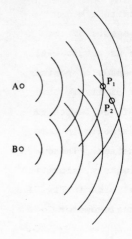

FIG. 1.7. Interference between coherent waves. The arcs indicate maxima of the amplitude at a given instant. At P_1 the disturbances reinforce at all times; at P_2 they cancel.

maximum and zero signal, and in fact this is the principle of several navigational systems. In optical terms the two sources generate an interference pattern. However, if the sources are in the optical rather than in the radio-frequency range, the phase relationship between the fields at a certain point is only constant for times comparable to the coherence time. Since the coherence time for ordinary sources is shorter than the response time for the fastest detectors, this means that it is impossible to observe interference between two different light sources. Alternatively, we can say that the interference field varies randomly more rapidly than the response time of the detector and so the detector records a time-averaged interference field. This is why optical interference experiments are always done with light from the same source, split and suitably recombined.

WAVES, PARTICLES AND RAYS

Theories of the nature of light have alternated between those involving waves and those involving particles. For waves the essential feature is the wavefront or surface of constant phase, and for particles it is the ray or particle trajectory. Very convincing experimental evidence is available for both kinds of

theories: interference and diffraction effects for wave
theories;   quantum phenomena of emission and absorption and
the effects of geometrical optics for particle theories.

The present-day view is that we do not understand everything
about light (or indeed about any other physical effect), and
that in order to obtain theories from which we can make true
predictions of experimental results we sometimes have to use
wave ideas and sometimes particle ideas.   This view has arisen
during the present century, from the development of quantum
mechanics.   According to this theory any particle has wave-like
aspects which must be appealed to in order to predict results
of some kinds of experiment, and equally, waves have particle-
like aspects which must be used in certain cases.   In any given
optical experiment either the wave or the particle aspect must
be stressed.

Fig.1.8 shows the relationships between some of the
principal different descriptions of light; beside each are

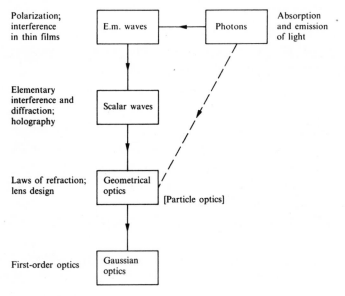

FIG.1.8.   Descriptions of light.

listed some of the effects for which the description is used.
We can regard these as a series of approximations, starting
from the most detailed (and mathematically most elaborate) in
terms of photons. Then if, loosely speaking, we allow the
Planck constant h to tend to zero we arrive at e.m. waves; if
we neglect all but one component of the electric vector we
obtain scalar waves; if we let the wavelength tend to zero we
get rays and geometrical optics; and finally, if we assume all
angles are very small, we obtain Gaussian or paraxial optics.
We must not, however, regard, say, Gaussian optics as being in
any sense 'wrong' because so many approximations were involved.
It is simply the right formulation for solving certain problems,
e.g. it is easy to get the well-known thin-lens formula

$$\frac{1}{l'} - \frac{1}{l} = \frac{1}{f'},$$

from Gaussian optics but it would be rather tedious to derive
it by rigorous quantum mechanics. In this way optics shows
clearly that we have no universal physical theory which will
explain and predict everything, and that meanwhile we must be
careful to use the appropriate theory for any given problem.

We can compare the wave concept of light with the wave
properties of fundamental particles such as electrons and protons.
In a picture at the same simple level which we have adopted
above in comparing the photon and e.m. wave pictures, the wave
associated with an electron has a complex amplitude which is
found as the solution of a certain partial differential
equation — the Schrödinger equation. The squared modulus of
this complex amplitude represents, to a suitable scale, the
probability of finding the electron in any chosen position, by
measurements made according to given rules. Thus the wave
nature of the electron, or other particle, appears as an
uncertainty in locating the particle, in accordance with the
Heisenberg uncertainty principle, and the waves associated

with material particles such as electrons and protons,
so-called *matter waves*, appear as distributions of probability
rather than field of force. In spite of the fact that matter
waves thus seem to have less of physical reality than e.m.
waves they nevertheless produce the standard wave-like effects
of interference and diffraction, and thus we can speak, for
example, of diffraction of neutrons and interference experiments
with electrons.

PROBLEMS

1.1. Calculate the wavelengths of e.m. waves of frequencies
10^6, 10^9, 10^{12}, and 10^{15}Hz.

1.2. Write down an expression for the complex amplitude of
a spherical wave diverging from a point source.

1.3. Calculate the power density in the spherical wave of
Problem 1.2, and show that this leads to the inverse-
square law.

1.4. Draw graphs of the energy in a photon as a function of
(a) wavelength and (b) frequency, using logarithmic
scales to accommodate the spectrum from X-rays to radio
waves.

1.5. What are the wavelength and frequency of radiation of
which the photon energy is of the order of magnitude of
the room-temperature thermal energy of atoms?

1.6. Plot a graph of coherence length against coherence time,
and mark on it points corresponding to typical sources
discussed in Chapter 1.

1.7. Calculate the quantum efficiencies at wavelength $0.45\mu m$
of the two photocathodes of Fig.1.5.

2. Geometrical optics

He faced about and, standing between the awnings, held out
his right arm at arm's length towards the sun. Wanted to try
that often. Yes; completely. The tip of his little finger
blotted out the sun's disc. Must be the focus where the rays
cross.

James Joyce, 'Ulysses'

THE USE OF GEOMETRICAL OPTICS

From the point of view of a pure physicist, geometrical optics
is a crude approximation for predicting in broad outlines, and
with many reservations, how e.m. waves behave. It can also
predict, with similar reservations, the trajectories of electrons,
neutrons, etc.

The applied physicist sees geometrical optics very differently.
It is his most important tool for designing many kinds of optical
system. Chiefly these are image-forming optical systems for
light and for electrons(e.g. optical and electron microscopes
and astronomical telescopes), but geometrical optics is essential
for some aspects of the design and use of almost any optical
system, from a shaving mirror to a single-lens reflex camera.
In addition, it is difficult to describe interference and
diffraction without using some of the ideas of geometrical
optics, such as mirrors and collimators.

The basic concept of geometrical optics is a simple, everyday
notion — light travels in a straight line unless it is reflected,
according to a law which seems intuitively obvious, or refracted,
according to a rather less obvious law. These laws can be
verified approximately using very simple apparatus, but the
accuracy with which elaborate optical systems work gives us a
very precise verification.

RAYS, WAVEFRONTS, REFLECTION, AND REFRACTION

A ray (of light) is a familiar concept, and we have to carry
out rather careful experiments to show that straight-line
propagation of light is not exactly true. In geometrical
optics we work in terms of rays of light and an associated
abstraction, the point source of light, giving (as in Fig.2.1)
a bundle or *pencil* of rays emitted in all directions. We admit
that light travels at a known finite velocity, so that in
Fig.2.1 we can mark on all the rays the points the light reaches
in a certain time t after leaving the point source P. If this

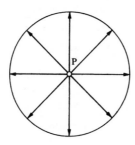

FIG.2.1. A geometrical wavefront as a surface of constant
phase.

is supposed to occur in a vacuum or — what is usually assumed
to be the same thing in geometrical optics — in air, these points
lie on a sphere of radius ct, as shown. If we reverted to a
wave picture of light, this sphere would be the surface reached
by a wave starting out from P at zero time, i.e. it would be
the wavefront, or the surface of constant phase. In the present
context we are not strictly concerned with waves, but these
surfaces of constant phase can be very useful in geometrical
optics, particularly if we consider them after the light has
passed through lenses or other optical components. Strictly
speaking they are called *geometrical wavefronts*, but we shall
call them simply wavefronts. Thus a wavefront is a surface
reached by the light from a point source in a certain time.

Clearly, the rays are normals to the wavefronts in Fig. 2.1. It happens that this is nearly always true in optical systems[†] although we shall not give the proof of this here, so that we can imagine a system of mutually orthogonal rays and wavefronts propagating through an optical system.

Rays of light go through lenses and mirrors according to the laws of *refraction* and *reflection*. The law of reflection, that the incident and reflected rays make equal angles with the normal to the reflecting surface and are coplanar with it, seems intuitively obvious on almost any theory of light. The origin of this law has not been traced; it was known to Euclid, in about 300 B.C. The law of refraction is also concerned with the relationships of the incident and refracted rays with the normal to the refracting surface, e.g. the surface of a sheet of glass. The two rays and the normal are again coplanar, which seems reasonable from symmetry, and the sines of the angles of incidence and refraction have a constant ratio, as in Fig.2.2[‡].

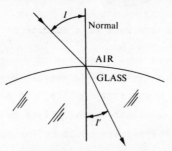

FIG.2.2. Snell's law of refraction.

This ratio depends on the materials on either side of the refracting boundary, and also on the wavelength. The law of refraction was discovered early in the seventeenth century by

[†]With the exception of certain crystals, but these are not usually used in the kind of optics considered in this chapter.

[‡]The law as stated is not true for certain crystals which are anisotropic, i.e. their optical and other properties vary with direction inside the crystal (see Chapter 4).

a Dutchman, W. Snell, and it is therefore called Snell's law.

At first sight the law of refraction seems obscure and
ad hoc. We may wonder, why sines rather than tangents, or any
other function, of the angles? However, in terms of wave
theory the form of the law is almost inevitable. Consider a
beam of parallel rays striking a plane refracting surface as
in Fig. 2.3, and let PP_1 be a wavefront of the incident beam

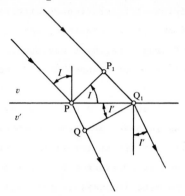

FIG. 2.3. Snell's law obtained from wave theory.

which meets the surface at P at time zero, say. After a
certain time t, P_1 has reached the surface at Q_1 and P has
travelled on to Q. Thus if the velocities in the two media
are v and v', we have

$$P_1 Q_1 = vt, \quad PQ = v't,$$

or

$$PQ_1 \sin I = vt, \quad PQ_1 \sin I' = v't,$$

from which we obtain by eliminating PQ_1,

$$\frac{\sin I'}{\sin I} = \frac{v'}{v}.$$

This is Snell's law. It is usual to put $v/c = 1/n$, where, as
in Chapter 1, c is the velocity in vacuum; n is then called
the *refractive index*. Snell's law now takes the form

$$n\sin I = n'\sin I'. \tag{2.1}$$

The refractive index of a material is a function of wavelength,

and for most transparent materials in the visible region it
lies between $\sim$ 1.3 and $\sim$ 2.3. The above argument leads also to
the law of reflection. At this point we can put them together and
for convenience, call them both Snell's law.[†]

Snell's law can be obtained in an entirely different way, from
Fermat's principle. This principle states that if light travels
from A to B through any optical system it will follow a path
such that the time of travel is stationary with respect to
neighbouring, but not physically possible, paths. 'Stationary'
means that the time of travel may be a maximum or a minimum,
or may simply have zero rate of change, as at a point of
inflection. The time function at a stationary point could also
be behaving like the altitude at the top of a mountain pass, a
minimum in some directions and a maximum in others (this is
called a saddle point). Pierre de Fermat first stated the
principle in 1657 in a form implying that the time of travel
is a *minimum* for the physically possible path ('Nature always
acts by the shortest course'), but stationarity is strictly
more correct. Since the velocity of light in a medium is
c/n, where n is the refractive index, this principle can be
stated in the form

$$\int_A^B nds \text{ is stationary,} \tag{2.2}$$

where ds is a differential element of length along any one of
the paths from A to B. This is illustrated in Fig. 2.4. The
integral of nds, which as we saw is proportional to the time of
travel of the light, is called the *optical path length*. Thus
the optical path length from a point source to all points on a
given wavefront is constant.

[†]This is not generally accepted usage, but it is very easy to
put (2.2) into the form of the law of reflection by putting
$n' = -n$, as a formal device, for a light ray returning into the
first medium after reflection. This then gives reflection as
a special case of refraction.

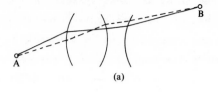

(a)

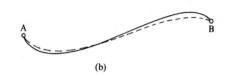

(b)

Fig. 2.4. Fermat's principle. The full line represents a physically possible ray path from A to B and the broken line another path. (a) For lenses. (b) For a medium of continuously varying refractive index.

Fermat's principle is analogous to the *principle of least action* proposed by Maupertuis (1744) as a foundation for mechanics. For a particle in a conservative field of force[†] the *action* is the integral of momentum along the trajectory, i.e. $\int p\,ds$, and the principle states that for this case the action for a physically possible path is a minimum. We can relate Fermat's principle and the principle of least action by formally making momentum proportional to refractive index. This can be justified for the photon model of light and also, with a suitable definition of refractive index, for electrons.

It is possible to derive Snell's law from Fermat's principle by finding the shortest optical path between points on either side of a refracting boundary (see Problem 2.3). Snell's law can be verified by experiments with prisms. The law is the

[†]In a conservative field the work done on a particle in transporting it from A to B is independent of the path from A to B. The force on the particle is the gradient of a scalar single-valued potential. Gravitational and electrostatic fields are examples of conservative fields, but a magnetic field is not conservative.

basis of optical design of all lens systems, and the fact that
they work as designed provides extensive experimental verification of the law and thus of Fermat's principle.

If we apply Snell's law (eqn (2.1) to a ray passing from
glass of refractive index n to air we find that at an angle of
incidence in the glass I_c, given by $\sin I_c = 1/n$, the refracted
ray 'emerges' at an angle of $90°$ to the normal; for larger
angles of incidence eqn (2.1) gives $\sin I' > 1$, and it is found
experimentally that the light is completely reflected at the
boundary — so-called *total internal reflection*. This follows
rather simply from the electromagnetic theory. The angle I_c is
called the *critical angle*.

OPTICAL IMAGES WITH A THIN LENS

The formation of images in the optical region by lenses
and mirrors is familiar, and it can be demonstrated in many
other regions of the e.m. spectrum. A simple explanation is as
follows. Fig. 2.5 shows a plano-convex lens, such as a
longsighted person might have in his spectacles. Light from a

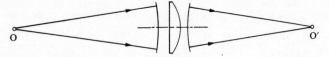

O   O'

Fig. 2.5. Formation of an image by a lens.

point source O on the axis, e.g. a pinhole in metal foil with
a lamp behind it, produces a diverging pencil of light with
spherical wavefronts convex to the lens. The lens is thicker
at the axis. Thus the optical path length through it is greatest
there, since the refractive index of the glass is greater than
that of air, and the wavefronts transmitted by the lens will be
retarded at the centre relative to the edge. Suppose for
simplicity that the emergent wavefronts are also spherical in
shape, as indeed they must be by symmetry for a small enough
diameter lens. Then they may be convex with a longer radius of

curvature, because of the greater delay at the centre, or, if
the shape of the lens and the position of O are suitable, they
may become concave, as indicated.   Drawing the rays as normals
to one of these wavefronts, we see that they intersect at some
point O´ on the axis — this is the image of O.

   If the object point O is not on the axis a similar argument
will show that an image point is again formed, with suitable
approximations.   Thus we get an image of an *extended object*.

   It remains to obtain simple formulae to describe image
formation, and this is most easily done by considering that
typically optical abstraction, the *thin lens*.   This is a lens
(as in Fig. 2.6) of a refractive index *n* and with curvatures

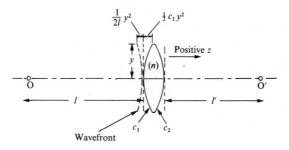

Fig. 2.6.   Formation of a point image.

(i.e. reciprocal of radius of curvature) c_1 and c_2 of the
refracting surfaces.   We neglect the thickness of the lens.   This
seems paradoxical for the lens in the figure, which is biconvex,
but the approximation gives useful results.   We assume that an
object point O at a distance l from the lens produces an image
at O´, a distance $l´$ from the lens, and we want to find $l´$ in
terms of l.

   We have to choose signs for these lengths, and to facilitate
this we set up a coordinate system with origin at the lens (it
does not matter exactly where in the lens since its thickness
is negligible) and z-axis along the optical axis.   Then *in our
figure* O´ has a positive z-coordinate so that $l´$ is a positive

number, and O has a negative coordinate, so that l is negative.
We can use the same system to give signs to the curvatures,
since the equation of a sphere passing through the origin and
with its centre to the right of the origin is of the form

$$z = \tfrac{1}{2}c(x^2 + y^2) + \ldots,$$

with positive c. Thus in our diagram it happens that c_1 is
positive and c_2 negative. We have now settled the problem of
signs according to the ordinary conventions of coordinate
geometry, and we can use the symbols as in coordinate geometry,
i.e. without further concern for signs.

At a distance y from the axis the lens will be $\tfrac{1}{2}(c_1 - c_2)y^2$
thinner than at the centre, by the above formula, i.e. this
amount of glass path is replaced by air, so that the optical
path through the lens between two planes tangent to the surfaces
will be shorter here than at the centre by $\tfrac{1}{2}(n-1)(c_1-c_2)y^2$.
The depth of curvature of the incoming wavefront is, as in the
figure, $y^2/2l$, and that of the emergent wavefront is $y^2/2l'$,
so that to ensure that optical path lengths (i.e. times of
travel between corresponding points of two wavefronts) are
equal we must have

$$\frac{1}{2l}y^2 + \frac{1}{2}(n - 1)(c_1 - c_2)y^2 = \frac{1}{2l'}y^2$$

or

$$\frac{1}{l'} - \frac{1}{l} = (n - 1)(c_1 - c_2). \tag{2.3}$$

This *conjugate distance equation* relates the positions of
object and image points, which are said to be *conjugates*. As
we have derived the result, the effect of the lens is to add
to the incoming wavefront an increment of curvature of amount
$(n - 1)(c_1 - c_2)$. If the incoming rays are parallel, i.e. the
incoming wavefronts are plane or the object point is at infinity,
the image distance l' is given by

$$l' = \frac{1}{(n - 1)(c_1 - c_2)},$$

as in Fig. 2.7. This distance is called the focal length f,

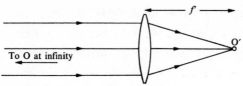

Fig. 2.7.   An object at infinity is imaged at a principal focus.

and P′ is the principal focus (any point where rays meet is a
focus). Similarly, if the object point is at a distance l given
by

$$l = \frac{-1}{(n - 1)(c_1 - c_2)} = -f,$$

the emergent rays are parallel, as in Fig. 2.8. Thus the point
at infinity and a principal focus are object and image
conjugates. These are two important cases for physical optics
applications. The second is used, as a *collimator*, to produce
a beam of plane waves from a point source; and the first is used
to show a diffraction or an interference effect which is
nominally formed at an infinite distance - in the *far field* -
at a convenient place for observation (see p.53).

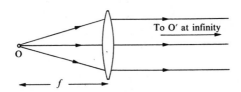

Fig. 2.8.   A collimator. The object at a principal focus
produces an image at infinity, i.e. a collimated beam.

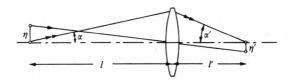

Fig. 2.9.   The image of an extended object.

If the lens forms an image of a small object, say a line segment of height η, as in Fig. 2.9, the image will have a height η´, and the ratio of these is the magnification m,

$$m = \eta´/\eta. \qquad (2.4)$$

Let α and $\alpha´$ be the semi-angles of the cones of rays forming the image, and let l and $l´$ be the conjugate distances. The ray from the end of the object through the centre of the lens passes through the lens undeviated, since the lens has its surfaces parallel at the centre, and so we have

$$\eta/l = \eta´/l´$$

or

$$m = l´/l. \qquad (2.5)$$

Also, if y is the height at which the other ray shown meets the lens,

$$\alpha = -y/l, \alpha´ = -y/l´,$$

so that, from (2.5),

$$m = \alpha/\alpha´. \qquad (2.6)$$

(2.5) and (2.6) are correct for all signs of the variables, as explained in the caption to Fig. 2.9. In all the above we have been tacitly using the Gaussian or paraxial approximation of small angles mentioned on p.18.

If the object is at the first principal focus, i.e. the focus on the object side, as in a collimator, the rays from each point of the object form a parallel pencil, $\alpha´$ is zero, and eqns (2.4) and (2.6) are not applicable. In this case we say that the image is formed at infinity and we use its angular subtense as a measure of its size. Thus if the object height is η (as above) and if the focal length is f, the ray from the end of the object through the centre of the lens emerges at an angle η/f to the axis, since it is undeviated, and all rays from this point therefore make this angle with the axis in the image space. Thus we have the rule that a point on the focal plane a distance η from the axis produces parallel wavefronts travelling at an angle

η/f to the axis (collimator). Conversely parallel wavefronts
travelling at an angle β to the axis form a point image at the
focal plane a distance βf from the axis (objective).

MULTI-ELEMENT LENSES

The above ideas are easily generalized to the lens systems
which contain a combination of several more-or-less thin lenses,
(e.g. the type of system used in a camera). From (2.5) and
(2.6) the magnification depends on the positions of object and
image, so for a multi-element or thick system we can find a
pair of conjugate planes with magnification unity. In most
practical cases these will be virtual conjugates, i.e. inside
the system as in Fig. 2.10, but this is not important. These
are known as the *principal planes*, and the axial points are the
principal points. Obviously, for a thin lens the principal points
coincide at the lens.

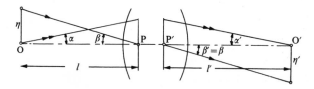

FIG. 2.10. The principal planes of a thick lens, P and P´.

Another ray from an axial object point 0 to its image 0´ passes
through the two principal planes at the same distance from the
axis, since they are planes of unit magnification. Now we can
compare Figs. 2.9 and 2.10 for a thin and a thick lens. The
figs. are similarly labelled to indicate their essential similarity,
but in the thick lens there is a kind of limbo or missing space
between the two principal planes. However, we can measure l from
the first principal plane to the object and $l´$ from the second
principal plane to the image, and all will correspond in the two
cases. For example, the focal length is the distance from the
principal plane to the point where rays from infinity focus

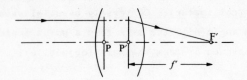

FIG. 2.11. The principal focus F' and the principal planes
for a thick lens.

(as in Fig. 2.11), and the magnification is again given by
$\alpha/\alpha' = l'/l$. Corresponding to (2.3), there will be a conjugate
distance equation,

$$\frac{1}{l'} - \frac{1}{l} = \frac{1}{f}, \qquad (2.7)$$

but now f, the focal length, depends on the detailed construction
of the lens system.

THE LAGRANGE INVARIANT AND THE POWER TRANSMITTED BY AN OPTICAL SYSTEM

We recall, from the previous two sections, the magnification
formula

$$m = \eta'/\eta = \alpha/\alpha',$$

where α and α' are the ray convergence angles and η and η' are
object and image heights. From this we have

$$\eta'\alpha' = \eta\alpha$$

as a relationship between object and image quantities. We can
generalize this to intermediate images formed between lenses in

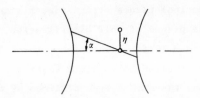

FIG. 2.12. The Lagrange invariant.

an optical system (as in Fig. 2.12), for the image formed by the
system to the left is the object for the next part, and they
share the same convergence angle α. Thus $\eta\alpha$ is the same in all

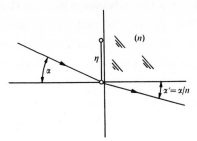

FIG. 2.13. An intermediate image formed at a plane surface.

air spaces in the system.

Now suppose a plane surface of glass of refractive index n is placed immediately after one of these intermediate images, as in Fig. 2.13. The image height is unaltered if the image is at the surface, but the convergence angle becomes $\alpha' = \alpha/n$ in Gaussian approximation. Thus finally we have that the quantity

$$H = n\alpha\eta \qquad (2.8)$$

is an invariant throughout an optical system. In this expression η is the intermediate image size corresponding to the original object size, α is the intermediate convergence angle of the ray from the original axial object point, and n is the refractive index. The quantity H is usually called the *Lagrange invariant*; it was discovered independently by several people including Lagrange.

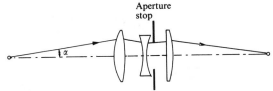

FIG.2.14. An aperture stop in a system determines the angle α of the accepted cone of rays.

As well as relating magnification and convergence angle, the Lagrange invariant is a measure of the light flux or power transmission capability of an optical system. In a real optical system the convergence angle α on the object side is usually

determined by the size of an iris diaphragm or aperture stop
in the system (as in Fig. 2.14), as well as, of course, the
distance of the object. Thus a small circular object of
radius η mm on the axis will radiate into the lens inside a cone
of half-angle α, and if the light power per unit area and per
unit solid angle (luminance) is B W mm^{-2}sr^{-1}, say, the power
collected by the lens will be $\pi^2 B\eta^2\alpha^2$ W. If we ignore attenuation
by reflection losses, absorption, and scattering, by conservation
of energy this same power flow must occur across any surface of
the lens system and across the final image surface. Thus on
comparing the above expression with (2.8) we see that the power
transmitted by the optical system is proportional to the square
of the Lagrange invariant.

The brightness or luminance of an image formed by an optical
system is the light power per unit area and per unit solid angle
in the image. It is clear from the Lagrange invariant that the
luminance of the image is equal to that of the object multiplied
by a factor less than unity which allows for attenuation, so that
no image can be brighter than the original object.

NON-PARAXIAL OPTICS

The results of pp.26 - 34 are all based on Gaussian optics,
according to which the lens and mirror apertures are supposed to
be very small and the rays all make small angles with the optical
axis. It is possible to define 'small' more precisely, as a
mathematical order of magnitude (Welford 1962), but here we shall
simply note that experimentally we do find well-defined images
under Gaussian conditions but if we go beyond a certain range of
angles the images look poorer (e.g. the image of a lamp formed
by a convex spectacle lens tipped obliquely, so that the object
and image are some way from the lens axis, is not sharp). Under
such conditions we cannot rely on the simple approximate equations
of this chapter and we have to trace rays exactly, i.e. according

to Snell's law. We then find that point objects do not form
point images, i.e. there are *aberrations*. Optical systems such
as camera lenses and microscopes have many lens components
arranged so as to correct or compensate aberrations.

Fig. 2.15 shows how one kind of aberration arises when we
use a thin convex lens at an angle θ to its axis. We saw on
p.28 that a thin lens forms an image of a point object by adding
an increment of curvature to the incident wavefront equal to
$1/f$, the reciprocal of the focal length;   this happens because
there is more glass at the centre than at the edge, so the
wavefront is retarded more at the centre.   In Fig. 2.15 this

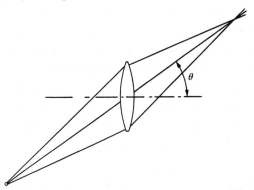

FIG. 2.15.   Oblique refraction by a thin lens. The rays in a
section at right-angles to the diagram focus at a greater distance
from the lens than those drawn.

effect will still happen in the section perpendicular to the
plane of the diagram, but in the plane of the diagram the width
of the lens presented to the wavefront is less by a factor
$\cos\theta$, and since there is, to a sufficient approximation, the same
variation in thickness between centre and edge, the increment in
curvature will be greater by a factor $1/\cos\theta$. Thus the refracted
wavefront will have a greater curvature in the plane of the
diagram than in the perpendicular section, and so the rays
(normals to the wavefront) will not focus to a single point.
This aberration is called *astigmatism*.

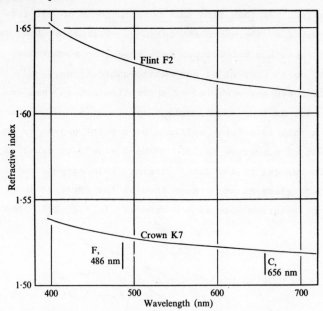

FIG. 2.16. Dispersion curves of two optical glasses. The names and symbols of the glasses have no physical significance.

Another example of an aberration is *chromatic aberration*. The refractive index depends on the wavelength of light, for all material media - an effect called *dispersion*. For example, Fig. 2.16 shows the refractive index as a function of wavelength for two common optical glasses. The brightest part of the visible spectrum lies between the two Fraunhofer spectrum lines C and F, and it can be seen that over this range the refractive index of the crown glass varies by about 0.008. Since the focal length of a thin lens (see p. 28) is given by $1/f = (n - 1)(c_1 - c_2)$, it can be seen that the focal length of a thin lens made of this glass would vary by about 0.008/0.52 = 1.6 per cent over this wavelength range. This effect is a form of chromatic aberration. Fortunately, as can be seen from the figure, the flint glass has about twice this variation of refractive index over the same wavelength range, and it is thus possible to combine a converging

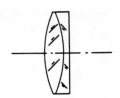

FIG. 2.17. An achromatic doublet.

lens of crown glass with a weaker diverging lens of flint glass, so as to cancel the chromatic aberrations but yet leave some converging power. Such a combination, as in Fig. 2.17, is an *achromatic doublet*.

PROBLEMS

2.1. Draw accurately rays refracted from air to glass ($n = 1.53$) at angles of incidence from 0° to 90°, at 10° intervals.

2.2. Points P_1 and P_2 are in media of refractive index n_1 and n_2 separated by a plane boundary. Calculate the optical length from P_1 to P_2 along straight-line segments from P_1 to a point Q on the surface and from Q to P_2. By differentiation prove Snell's law.

2.3. Sketch typical object and image positions, indicating suitable rays, for (a) a camera, (b) a slide projector, and (c) a burning glass.

2.4. Show that the optical path length along all rays between two wavefronts of a pencil is constant.

2.5. A surface of curvature c separates media of refractive indices n and n'. Show that the conjugate distance equation for this surface is

$$\frac{n'}{l'} - \frac{n}{l} = (n' - n)c.$$

2.6.   From p.28 show that the conjugate distance equation for
a thin lens can be written

$$\frac{1}{l^{\prime}} - \frac{1}{l} = \frac{1}{f} \; .$$

Plot a graph of $l^{\prime}$ as a function of l for a lens of
focal length 100 mm, allowing the conjugates to range
between − 1000 mm and + 1000 mm.

2.7.   What is the physical significance of positive values of
l and negative values of $l^{\prime}$ in the above example?

2.8.   Find a conjugate distance equation for a concave mirror.

2.9.   A thin lens has focal length 50 mm. What is the
magnification for the following object conjugates:
20 mm, − 100 mm, + 100 mm.

2.10. What is the focal length of a concave mirror of radius
r? Calculate and draw to scale the image positions and
magnifications for a concave mirror of radius 100 mm for
object distances 25 mm, 100 mm, 200 mm.

2.11. A lamp filament in the form of a flat ribbon of area
10 mm^2 radiates 10W of light. (a) What is its luminance,
and (b) how much power is collected by a lens of diameter
30 mm which is 100 mm from the filament?

2.12. Show that a prism with small angle α deviates light rays
through an angle $(n - 1)\alpha$, where n is the refractive index
of the prism.

3. Propagation of waves: interference and diffraction

Be a warm day I fancy. Specially in these black clothes feel
it more. Black conducts, reflects (refracts is it?) the heat.

James Joyce, Ulysses

In the geometrical optics approximation we made the tacit
assumption that rays intersect without interacting with each
other. This would mean that rays meeting at a point, as at the
image of a point source formed by an aberration-free lens,
produce an infinitely small point image; this is known
experimentally to be untrue. It would also mean that, if two
beams of light overlap on a screen, the resultant light intensity
(power density) is the sum of the intensities in the individual
beams; experimentally this is sometimes true, sometimes false.
In this chapter we examine these effects, which are, of course,
examples of *diffraction* and *interference*.

INTERFERENCE OF TWO BEAMS

We saw in Chapter 1 that nominally monochromatic light beams
have very rapid random phase variations. Thus in order to see
interference effects between two beams we must ensure that these
phase variations are the same and in step in both beams. This is
done by taking both beams from the same light source. It is
simplest to think first about beams of plane waves intersecting
at an angle θ. There were many classical experiments in which
this was done in different ways. Fig. 3.1 shows one way in which
it might be done with modern equipment. If the region where the
beams intersect is examined, e.g. by putting a white screen there
or by scanning a small photodetector across it, straight dark and
light bands perpendicular to the plane of the diagram are found,
i.e. *interference fringes. Bright fringes are formed whenever
the two waves are in phase*. The inset to Fig. 3.1 shows

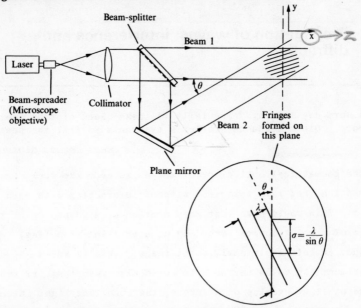

FIG. 3.1. A convenient way to form interference fringes. Two
beams of collimated light from the laser intersect at an angle
θ, and fringes are formed where the beams cross. The beam-
splitter is a plate of glass with a thin semi-transparent film
of aluminium on one surface.

wavefronts from the two beams at a given instant, and from this
it can be seen that the spacing σ between the fringes is given by

$$\sigma = \lambda/\sin\theta, \tag{3.1}$$

since it corresponds to the intersection of one wavefront of
beam 1 by successive wavefronts of beam 2.

In slightly more detail, if we suppose the beams are of equal
intensity we can represent their complex amplitudes by (p.7),

$$\left.\begin{array}{l} \text{beam 1: } E\exp\{-2\pi i z/\lambda\}, \\[2mm] \text{beam 2: } E\exp\{-2\pi i(z\cos\theta + y\sin\theta)/\lambda\}. \end{array}\right\} \tag{3.2}$$

The total complex amplitude in the interference pattern across
the plane $z = 0$, which we can take to be the plane of observation,
is then

$$E\{1 + \exp(-2\pi i(y/\lambda)\sin\theta)\},$$

put z=0 into eqⁿ 3.2 & add exponents

$E + E\exp\left\{-2\pi i y\sin\theta/\lambda\right\}$

$= E\left\{1 + \exp\left(-2\pi i(y/\lambda)\sin\theta\right)\right\}$

and the observed intensity is the squared modulus of this, i.e.

$$I(y) = 2E\{1 + \cos(\frac{2\pi}{\lambda} y\sin\theta)\}$$

$$= 4E\cos^2(\frac{\pi}{\lambda} y\sin\theta),$$

$1 + \cos 2\varphi$ (3.3)
$= 2\cos^2\varphi$

This function, giving what are usually called $\cos^2$ fringes,
is plotted in Fig. 3.2. Since each time the argument of the
cosine increases by π we go through a complete period of the
fringes, we have verified (3.1) above for the fringe spacing.
The $\cos^2$ light-intensity distribution can be verified by photoelectric scanning, but visually the fringes appear to have much
narrower dark regions than in the figure. This is a consequence

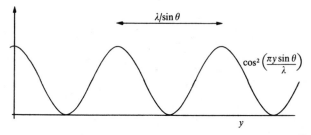

Fig. 3.2. The light-intensity distribution across $\cos^2$ fringes.

of the very nonlinear response of the eye (p.11), and it is often
misleading in making a quick visual assessment of an interference
or diffraction effect.

If the beams are of different complex amplitudes, and
therefore of different intensities, the minima in the fringe
system will not be zeros, i.e. there will not be maximum contrast
or visibility in the fringes (see Problem 3.3).

Generally, we are mainly interested in the fringe spacing
and contrast rather than in the details of the intensity
variation in the fringes. The maxima occur where the two beams
are in phase. We can generalize this immediately by noting, from
Chapter 2, that points of equal phase occur where the optical
path lengths from the source via the two interfering beams to the

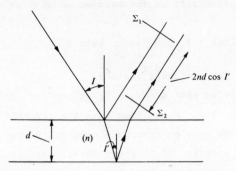

FIG. 3.3. The difference of optical paths between beams reflected at the first and second surfaces of a layer of thickness d and refractive index n.

point in question are the same, or where they differ by a whole number of vacuum wavelengths. This is then applicable to interference in media of different refractive index. A good example is the oil film on water or, more generally, a layer of index n and thickness d, as in Fig. 3.3. Let a collimated (parallel) beam of wavelength λ meet the layer at an angle of incidence I. Some light is reflected at each surface, and there will be a path difference between corresponding wavefronts. This is indicated in the figure where Σ_1 and Σ_2 have originated from the same wavefront after a certain time. It is a simple exercise in the use of Snell's law (Chapter 2) to show that the optical path difference between the two beams is

$$2nd\cos I', \tag{3.4}$$

where I' is the angle of incidence *inside* the layer. Thus a bright fringe is formed whenever

$$2nd\cos I' = N\lambda, \tag{3.5}$$

where N is an integer, the *order of interference*. (In certain cases there is a phase change on reflection at one surface, and we have $(N + \frac{1}{2})\lambda$ on the right-hand side.)

Eqn (3.5) shows how variations in the thickness d of an oil film are indicated by the shape of the interference fringes. Similarly,

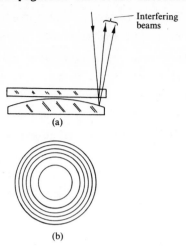

(a)

(b)

FIG. 3.4. Newton's rings formed by interference between beams reflected at a curved interface. (a) The apparatus. (b) The spacing to scale of successive bright fringes.

in the classical experiment of Newton's rings (Fig. 3.4) the successive fringe diameters indicates the variation of the thickness of the gap between the two glass surfaces. This equation also has other applications, as will be seen in the next section and in Chapter 6.

INTERFERENCE WITH EXTENDED AND POLYCHROMATIC LIGHT SOURCES.

In discussing interference between the beams reflected from an approximately parallel layer in the previous section we assumed the incident light was monochromatic and collimated over a reasonable area of the layer, and this implied that it came from a single point source at a great distance. However, we normally see such effects under less stringent conditions: the source may cover a large extent, e.g. the sky for oil films or a sodium lamp for Newton's rings, and it may have a range of different wavelengths in it. We can understand this by referring again to (3.5).

First, consider an extended source. Each source point will form its own interference fringes, and these fringe systems will be

independent, i.e. the intensities will add. The angle of
incidence I' will be different for different source points, so
that the path difference (eqn(3.4)) at a given part of the
layer will vary, and thus the fringe maxima will not coincide.
However, if the thickness d is small enough quite a large
variation in the angle of incidence is needed to change the path
difference by, say, $\lambda/4$, so that the fringe systems will all
more-or-less coincide, and fringes can be seen with an extended
source. Problem 3.5 illustrates this. In Chapter 1 we spoke
of the need to have *coherence* between light beams if they are to
show interference effects. In the present case we see that the
source has to be restricted in size (more exactly in angular
subtense) to ensure that the beams reflected from the two surfaces
of the layer are coherent.

If we now put eqn (3.5) in the form

$$\frac{2nd\cos I'}{\lambda} = N, \tag{3.5a}$$

we see that for given thickness and angle of incidence the order
of interference N will vary with the wavelength. If N is non-
integral it is interpreted as the number of wavelengths, possibly
fractional, of path difference between the interfering beams.
Thus if the source is polychromatic the fringe systems from the

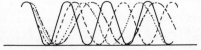

FIG. 3.5. Superimposition of intensities of fringes formed in
polychromatic light.

various wavelengths will again be displaced and will add in
intensity to give a more-or-less uniform appearance, as in
Fig. 3.5. However, again if the layer is thin enough, i.e. d is
small enough, N in eqn (3.5a) will vary very little over a
reasonable wavelength range, and fringes of good contrast will
be obtained. Thus for coherence we have to restrict the
wavelength range.

Two-beam interference effects are used in many devices. A few of these devices are discussed in Chapter 6, and others are described in classical texts (e.g. Ditchburn 1952; Longhurst 1973). From the above we can make a generalization which applies with appropriate modifications to all of these.

Two-beam interference effects can be obtained with polychromatic extended light sources. The contrast or visibility of the fringes depends on both the bandwidth (frequency or wavelength spread) and the angular extent of the source. Generally both of these must decrease with increasing optical path difference between the beams if the visibility is to be kept high. In Chapter 6 we shall see applications of this principle to astronomy and spectroscopy.

DIFFRACTION

In the previous section we started by discussing interference between collimated beams, as in Fig. 3.1. The beams were regarded as composed of plane waves of indefinitely great width - i.e. simply as described by, say, eqn (1.5) - with no restriction placed on the position vector which indicates the point in space at which we consider the wave disturbance. In fact the beams are limited in extent by the diameters of the collimator lenses in Fig. 3.1, and these certainly cannot be considered as indefinitely large. This restriction does not materially affect the description of two-beam interference; however when we examine the propagation of a single collimated beam from a point source we find that it does not propagate indefinitely with a sharply defined rim given by the edge of the lens; instead, the disturbance spreads out and becomes uneven near the edge in a complicated way. This is indicated in Fig. 3.6, which shows the light intensity observed in line with the edge at different distances. We now discuss this effect, known as diffraction.

We can simplify the discussion by considering first a

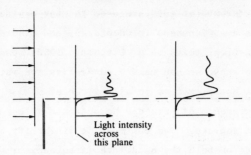

FIG. 3.6.   Diffraction at an edge.   The graphs show the light
intensity near the geometrical shadow at two distances from the
edge in the ratio 1:4.

collimated beam of large diameter which meets an opaque straight
edge: experimentally the effect is very much as in Fig. 3.6,
i.e. it does not much matter if the edge is curved or straight.
Similar effects are found in the propagation of all waves;   in
sound and water waves it is easy to observe diffraction at an
edge or any obstacle because the wavelength is large.   Quali-
tatively we can see why diffraction occurs (i.e. why there is not
a sharp shadow at all distances from the edge in Fig. 3.6) in
terms of a basic physical principle that discontinuities do not
occur in the wave representation.   Thus the wave disturbance
cannot stop abruptly at the line of the geometrical shadow, but
must decay gradually.   However, this does not give us a
quantitative picture.

   It can be shown (Chapter 10 of *Electromagnetism*(OPS 1)) that
in a uniform medium the electric and magnetic fields of e.m.
radiation both obey a partial differential equation - the
equation of wave motion.   Thus for one component of the electric
field vector, E_x say, we have, in a uniform dielectric medium

$$\nabla^2 E_x = \mu_r \mu_0 \varepsilon_r \varepsilon_0 \ddot{E}_x, \qquad (3.6)$$

where μ_0 and ε_0 are the permeability and permittivity of vacuum
and μ_r and ε_r are the relative permeability and the relative
permittivity of the medium.   For monochromatic (single-frequency)

waves we can put $E_x = E\exp i\omega t$ in eqn (3.6), where E now becomes
a complex amplitude (Chapter 1), and we obtain

$$\nabla^2 E + \omega^2 \mu_r \mu_0 \varepsilon_r \varepsilon_0 E = 0, \qquad (3.7)$$

the time-independent equation of wave motion.

   To solve the diffraction problem in complete generality for
e.m. waves we should have to solve eqn (3.7) and five others
like it for all components of $\underline{E}$ and $\underline{H}$, using appropriate
distributions of μ_r and ε_r and putting in suitable boundary
conditions at the diffracting obstacles. This has been done for
some simple cases, and the results have been verified experimentally by measurements with microwaves ($\lambda \sim 10$ mm). In the
optical region we can greatly simplify the problem by considering
just one component of $\underline{E}$ (the scalar wave theory) and by using
some approximations which are good for most regions of practical
interest. Roughly, these regions are those at a large distance
from the diffracting structures, where 'large' means many
wavelengths and where the diffracting angles are small, i.e. less
than, say, 0.1 rad. Fig. 3.7 illustrates these regions for
diffraction of a wave at an aperture in a screen.

   We then have the following physical picture. In wave
propagation, as the disturbance progresses through the medium,
each point reached by the disturbance becomes in turn a starting

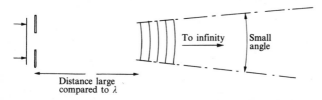

FIG. 3.7. The region in which scalar diffraction theory can be
safely applied.

point for the disturbance which moves on from that point. This
is easily pictured in the case of a transverse disturbance
travelling along a stretched string or for water waves from an

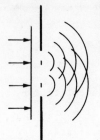

FIG. 3.8.   Huygens' secondary wavelets.   Two secondary sources are indicated in the aperture.

origin.   For diffraction of a plane wave at an aperture in a screen we proceed as in Fig. 3.8.   The wave travels from left to right and meets a plane opaque screen with an aperture in it. To find out what happens on the right of the screen we assume that from each point in the plane of the screen secondary spherical wavelets spread out.   The resultant disturbance is to be obtained by adding the complex amplitudes of all of these, *taking account of their relative phases*.   Thus we consider interference between different elements of the same disturbance in the plane of the screen.   If the screen were not there we should expect this curious procedure to give the same result as if the plane waves were simply carried on, and this is found to be so.

These ideas were roughly formulated by Huygens in the seventeenth century, refined by Fresnel early in the nineteenth century, and given a definite mathematical form by Kirchhoff

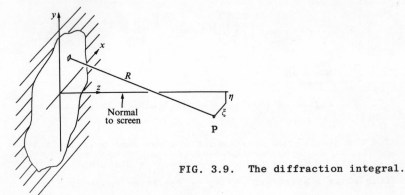

FIG. 3.9.   The diffraction integral.

about 80 years later. We shall give here a special case which
is simpler than the general formulation but which applies to
most problems of current interest.

We take a rectangular coordinate system as in Fig. 3.9 with
the x,y-plane as the plane of the screen, so that the arriving
wavefronts are parallel to this plane, and we wish to determine
the complex amplitude at a point P with coordinates (ξ,η,ζ).
According to the above ideas, we assume that an element $dxdy$ of
the wavefront in the aperture acts as a source of a secondary
spherical wavelet of strength proportional to the area of the
element. Thus the complex amplitude at P due to this element is

$$\frac{E_0}{\lambda} \cdot \frac{\exp(-ikR)}{R} dxdy,$$

where $k = 2\pi/\lambda$ as on p.6. E_0 is a constant complex amplitude.
The factor E_0/λ occurs naturally in the development according to
Kirchhoff. Here we shall accept the factor as a way of keeping
the equations dimensionally correct. The total effect at P due
to all the incident wave which passes through the aperture is
then

$$E_P = \frac{E_0}{\lambda} \iint \frac{1}{R} \exp(-ikR)\,dxdy. \tag{3.8}$$

The integral is to be taken over the whole of the aperture and
R is, of course, a function of x and y as different elements of
the wavefront are taken.

To express R in a manageable way, we have, from Pythagoras'
theorem,

$$R^2 = (x - \xi)^2 + (y - \eta)^2 + \zeta^2$$
$$= x^2 + y^2 - 2(\xi x + \eta y) + \xi^2 + \eta^2 + \zeta^2.$$

Thus

$$R = \xi \left\{ 1 - \frac{2(\xi x + \eta y)}{\zeta^2} + \frac{x^2 + y^2}{\zeta^2} + \frac{\xi^2 + \eta^2}{\zeta^2} \right\}^{\frac{1}{2}}$$

ζ ?

$$= \zeta - \frac{\xi x + \eta y}{\zeta} + \frac{x^2 + y^2}{2\zeta} + \frac{\xi^2 + \eta^2}{2\zeta} + \ldots, \tag{3.9}$$

on expanding as far as the first term by the binomial theorem. As we should expect, R consists of a term $\zeta + (\xi^2 + \eta^2)/\zeta$ which is independent of x and y and which is large in terms of λ, together with some smaller terms. Whenever R increases or decreases by one wavelength (as a consequence of varying x and y in the integration) the exponent in eqn (3.8) changes by 2π, and the complex amplitude in the integrand goes through a complete cycle. Thus small changes in R are important in the exponential. On the other hand, since we are supposing R is large compared with λ, such changes can be ignored in the $1/R$ factor, and this can be written $1/\zeta$ and taken outside the integral. Since we are assuming small diffraction angles we need only consider linear terms in ξ/ζ and η/ζ in eqn (3.9). Eqn (3.8) then becomes

$$E_p = \frac{E_0}{\lambda \zeta} \exp(-ik\zeta) \times$$

$$\times \iint \exp\left\{\frac{ik}{\zeta}(\xi x + \eta y) - \frac{ik}{2\zeta}(x^2 + y^2)\right\} dx dy. \tag{3.10}$$

We now make an important simplifying assumption, that ζ is so large that the term in $x^2 + y^2$ can be neglected. This means that the maximum value of $(x^2 + y^2)/\zeta$ anywhere in the aperture is much less than λ. The physical implication will be seen in the next section. The factor $\exp(-ik\zeta)$ is a constant which will give a factor of unity when we take the squared modulus of E_p to get the light intensity, so it can be dropped, and we have as the final

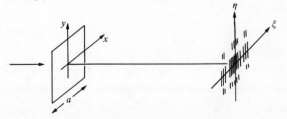

FIG. 3.10. Far-field diffraction at a square aperture.

expression for the complex amplitude at P,

$$E_P = \frac{E_0}{\lambda\zeta} \iint \exp\left\{\frac{i2\pi}{\lambda} (\xi x + \eta y)\right\} dxdy. \qquad (3.11)$$

We can apply this immediately to the simple but useful case of a square aperture of side a, as in Fig. 3.10. We take the origin at the centre of the square, and we let ξ be the z-coordinate of the plane in which we want to find the diffraction pattern. The double integration in eqn (3.11) splits into two factors,

$$E(\xi,\eta) = \frac{E_0}{\lambda\zeta} \int_{-\frac{1}{2}a}^{\frac{1}{2}a} \exp\left(\frac{i2\pi}{\lambda\zeta}\xi x\right) dx \int_{-\frac{1}{2}a}^{\frac{1}{2}a} \exp\left(\frac{i2\pi}{\lambda\zeta}\eta y\right) dy$$

$$= \frac{a^2 E_0}{\lambda\zeta}\left\{\frac{\sin(\pi a\xi/\lambda\zeta)}{\pi a\xi/\lambda\zeta}\right\} \cdot \left\{\frac{\sin(\pi a\eta/\lambda\zeta)}{\pi a\eta/\lambda\zeta}\right\} .$$

The notation sinc x is used for $(\sin\pi x)/\pi x$, so that the diffracted complex amplitude at (ξ,η) is

$$E(\xi,\eta) = \frac{a^2 E_0}{\lambda\zeta} \text{ sinc}\left(\frac{a\xi}{\lambda\zeta}\right) \text{ sinc}\left(\frac{a\eta}{\lambda\zeta}\right). \qquad (3.12)$$

The light intensity $I(\xi,\eta)$ is the squared modulus of the complex amplitude (p.7), so we have

$$I(\xi,\eta) = \frac{a^4 E_0^2}{\lambda^2\zeta^2} \text{ sinc}^2\left(\frac{a\xi}{\lambda\zeta}\right) \text{ sinc}^2\left(\frac{a\eta}{\lambda\zeta}\right). \qquad (3.13)$$

The general form of this pattern, a central maximum of intensity with surrounding subsidiary maxima and minima, is indicated in Fig. 3.10 by the hatching (see also Problem 3.7). In describing such patterns we usually rescale the intensity by normalizing it to unity at the centre of the pattern, i.e. at $\xi = \eta = 0$. For eqn (3.13) this simply means omitting the factor $a^4 E_0^2/\lambda^2\zeta^2$; this is acceptable for most problems in physical optics, but the physical dimensions are lost. The factor $a^4 E_0^2/\lambda^2\zeta^2$ indicates that the central intensity is proportional to the fourth power of the linear dimensions of the aperture and inversely proportional to the square of the wavelength. These are general rules applying to all diffraction of this kind, where quadratic terms in the aperture are negligible. From the argument of the sinc function in eqn (3.13) the lateral scale, i.e. distances between successive maxima, varies inversely as the aperture size and directly as the wavelength, and these are again general rules.

We stress that the results given in this section are obtained after a succession of approximations from the original formulation of Kirchhoff, and that the latter itself is only an approximate way of solving the six partial differential equations, like eqn (3.7), obtained from Maxwell's equations. Again Maxwell's equations would be inadequate to represent diffraction under some conditions of very high or very low light intensities. We justify the use of eqn (3.11) by the fact that experimentally it is verified to good accuracy in most ordinary situations.

DIFFRACTION IN THE FAR FIELD

Suppose that in Fig. 3.9 we observe the diffracting aperture from the point P at which the diffracted field is being found and suppose also that we could see the variations in complex amplitude in the aperture due to the variation in optical path length R from P to points (x, y) in the aperture (this could be done by for example, a suitably arranged interferometer). The assumption we made that the term in $x^2 + y^2$ in eqn (3.10) is negligible means that ζ is taken so large that this phase variation over the aperture would be linear rather than quadratic, since it is in fact the exponent in eqn (3.10). The distance ζ from the diffracting aperture then satisfies

$$\zeta \gg a^2/\lambda, \tag{3.14}$$

and we have *far-field* diffraction or *Fraunhofer* diffraction. If on the other hand the quadratic term is not negligible then

$$\zeta < a^2/\lambda, \tag{3.15}$$

and we have *near-field* or *Fresnel* diffraction.

We shall consider only far-field diffraction. In the limit as ζ gets very large we can put $\xi/\zeta = u, \eta/\zeta = v$, where u and v are angular coordinates, and we speak of light diffracted into a direction (u,v). Eqn (3.13) is then written in normalized form

$$I(u,v) = \text{sinc}^2\left(\frac{au}{\lambda}\right)\text{sinc}^2\left(\frac{av}{\lambda}\right) \tag{3.13a}$$

We do not actually have to go to a distance given by eqn (3.14) to observe far-field diffraction. From p.30 parallel rays in a direction with components (u,v) on one side of a lens come to a focus at a point with coordinates (fu,fv) on the focal plane of the lens. Thus an objective (collimator in reverse) can be used to bring the far field to a convenient distance, as in Fig. 3.11.

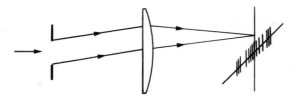

FIG. 3.11. Bringing the far field to a convenient place for observation. The far-field diffraction pattern of the aperture is formed at the focal plane of the lens. (If in addition we want the phase relationships between different parts of the far-field pattern to represent correctly the Fourier transform of the aperture we must also put the aperture at the front focal plane of the lens.)

We can re-write eqn (3.11) normalized to have unity intensity at the centre of the far-field, i.e. in the direction of the incident wave, given by $\xi = \eta = 0$, and we can also write the far-field coordinates as the diffraction angles (u,v). We then have for the normalized complex amplitude at (u,v),

$$E(u,v) = \frac{1}{A}\iint\exp\left\{\frac{i2\pi}{\lambda}(ux + vy)\right\}dxdy, \tag{3.11a}$$

where A is the area of the diffracting aperture and, as before, the integration is over the aperture. Next we specify the diffracting aperture by means of a function $F(x,y)$, which is put as a factor inside the integral and which is defined as equal to unity for (x,y) inside the pupil and zero outside. This enables us to formally extend the limits of integration to infinity and eqn (3.11a) becomes

$$E(u,v) = \frac{1}{A}\int\!\!\!\int_{-\infty}^{\infty} F(x,y)\exp\left\{\frac{i2\pi}{\lambda}(ux + vy)\right\}\,dxdy. \quad (3.11b)$$

The introduction of the function $F(x,y)$ is more than a mere formal device. It need not be only a binary function (i.e. taking only values 0 or 1); it can be modified to give the effect of a screen across the aperture which absorbs some light or which has a phase-changing effect. Both these devices are useful; if there is absorption the normalizing factor strictly has to be interpreted as $(\int\!\int F(x,y)\,dxdy)^{-1}$.

Eqn (3.11a) can be interpreted as a Fourier transform relationship (see Appendix). We see that if eqn (3.11b) is rewritten in the form

$$f(s,t) = \int\!\!\!\int_{-\infty}^{\infty} \frac{F(x,y)}{A}\exp\left\{i2\pi(sx + ty)\right\}\,dxdy \quad (3.11c)$$

where we have put $s = u/\lambda, t = v/\lambda$ as new variables and where $E(\lambda s,\lambda t) \equiv f(s,t)$, we can say that the complex amplitude in the far-field diffraction pattern is the (inverse) Fourier transform of the complex amplitude in the diffracting aperture, to a suitably chosen scale.†

We can see from eqn (3.11a) that the size of the far-field pattern is proportional to the wavelength and inversely pro-

†The question of whether it is the inverse or direct transform is not physically significant. The sign of the imaginary unit (i.e. ±i) in the complex amplitude can be chosen arbitrarily in the first instance, since we are only dealing with the real part of the complex amplitude.

portional to the linear scale of the diffracting aperture; this
is, from the general properties of Fourier transforms, (Appendix)
true for any shape of aperture, in agreement with the rule stated
in the previous section.

We meet far-field diffraction in many different physical
situations. A microwave paraboloid antenna has a characteristic
angular distribution of radiated power which is the far-field
diffraction pattern of the 'aperture' formed by the rim of the
Paraboloidal reflector. If we could avoid certain distorting
effects due to atmospheric turbulence, the image of a star formed
by a telescope would be the far-field diffraction pattern of the
aperture of the telescope mirror. The Laue X-ray diffraction
pattern from a single crystal is the far-field diffraction
pattern of a regular lattice of diffracting points, i.e. it is the
Fourier transform of a periodic array of delta functions.

The far-field pattern for a circular aperture must be radially
symmetrical about the axis of the aperture. Taking only one
angular coordinate u, which is the (small) angle between the
axis and the direction in the far-field in which we are interested,
it is found that the complex amplitude in the farfield is in
normalized form

$$E(u) = \frac{2J_1(2\pi au/\lambda)}{2\pi au/\lambda} , \qquad (3.16)$$

where $J_1(z)$ is the Bessel function of the first kind and first
order. This function, which is available from many books of
tables, behaves like an attenuated sine wave.† The intensity in
the diffraction pattern is plotted in Fig. 3.12 with logarithmic
ordinate scale. It is known as the *Airy diffraction pattern,* or
Airy disc, after the astronomer G.B. Airy, who calculated it as
the theoretical form of a star image.

†A comprehensive collection is *Handbook of mathematical
functions with formulas, graphs and mathematical tables* by M.
Abramovitz and I. Stegun (National Bureau of Standards,
Washington, D.C., 1965).

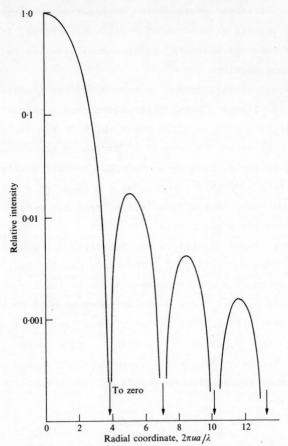

FIG. 3.12. The Airy pattern, or light-intensity distribution in the image of a point source of monochromatic light formed by a system with a circular aperture. The function plotted is the squared modulus of the amplitude, as in eqn (3.16).

We can observe far-field diffraction effects in the optical region most easily with a laser as light source. It is necessary that all parts of the wavefront in the diffracting aperture should be able to interfere with each other, i.e. they must be coherent with each other in the sense explained on p.16. The light from a helium-neon laser in correct adjustment behaves as if it came from a single point source, and it is therefore coherent over the whole

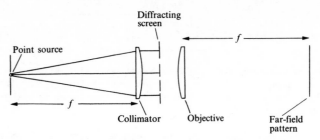

FIG. 3.13. A diffractometer. The far-field pattern of the screen is formed at the focal plane of the objective.

wavefront. Fig. 3.13 shows a typical arrangement of apparatus for producing far-field diffraction patterns; diffracting screens of different shapes are placed in the collimated beam. Many beautiful examples of far-field diffraction patterns are given elsewhere (e.g. Lipson 1972).

It is often useful to estimate the general features of a far-field diffraction pattern without a detailed calculation. It is usually true that if the diffracting aperture is simple in shape and has no blanked-off area in the middle then the maximum intensity in the farfield is in the direction of the incident wavefront. If we go away from this direction by an angle λ/d, where d is a distance of the order of magnitude of the width of the aperture, this will be roughly the direction of the first minimum. Thus the angular half-width of the central maximum, i.e. the full width of the pattern at half maximum intensity, is of order λ/d. For example, a laser beam about 1 mm in diameter (i.e. as it comes from the laser) will spread by diffraction, even if nominally collimated, over an angle of about 1 mrad, but if the beam is first spread out by means of a beam-expander (as in Fig. 3.14) to, say, 20 mm, it will only spread at about 0.05 mrad.

INTERFERENCE, DIFFRACTION, AND THE PHOTON PICTURE

Interference and diffraction are now thought of as essentially

FIG. 3.14. A beam-expander. This is an afocal system, i.e.
it forms an image at infinity of an object at infinity. The
incoming beam is expanded in the ratio of the focal lengths of
the two lenses.

wave phenomena, although there have been attempts (e.g. by Newton)
to explain interference using a particle model. Yet the dual
nature of light - wave and particulate - is well established, and
the apparent contradiction at an elementary level between these
aspects is perhaps more striking than for any other particle-
cum-wave. The contradiction is sometimes handled as on pp.16
-19 simply by saying that we have to use the wave representation
for some purposes and the particle representation for others,
but it is possible to enlarge on this approach.

Suppose we have an interferometer, such as in Young's
experiment (p.91), and we wish to discuss its properties in
terms of photons rather than waves. Then according to the
methods of quantum mechanics we have to consider the passage of
a single photon at a time through the apparatus. Indeed, with
an ordinary light source and an apparatus of reasonable size,
it can easily be shown that it is very unlikely that more than
a single photon will be in transit through the apparatus at any
given time. Thus we imagine a detector, in the plane where the
fringes are to be formed, which builds up the fringe pattern as
dots, one for each arriving photon in the position where it
activates the detector. This experiment was first done by
G.I. Taylor using photographic plates, and it has since been
repeated in many different ways. The results always show that
the photons appear at first to be arriving at random positions,
but gradually, as more photons arrive, the classical interference
pattern as predicted by wave theory is built up.

We explain this by saying that the photon does not follow

a definite path through the apparatus, but that it can follow
any of several different paths. Clearly there should be a high
probability for the photon to follow a path terminating near the
position of an interference fringe maximum, and a low probability
for it to arrive near a minimum. To calculate these
probabilities we should have to solve Schrödinger's equation,
putting in as boundary and initial conditions the shapes and
sizes of the slits, etc. in the apparatus and the momentum and
point of origin of the photons. If we were to do this we should
find that we arrived eventually at the same form for the fringe
system as that obtained by elementary wave theory. Thus we find
the probabilities for the arrival of photons giving the same
interference or diffraction patterns as the intensity
distributions of interfering or diffracted waves.

In the above hypothetical calculation, although all the details
of the apparatus have to go into the equations, all we get out
at the end is a probability density for photons to be detected in
a certain region. Thus we have by-passed the tantalizing question,
which of the two slits in the interferometer did a certain photon
go through? According to the rules of quantum mechanics we are
not permitted to ask this question. We may only ask questions
whose answers can be determined experimentally without upsetting
the outcome of the main experiment. If we were to attempt to
find out which slit the photon went through, we should have to
annihilate it in the process of detecting it at one or other slit.
Thus it would then not take part in forming the interference
pattern which was the original object of the experiment.

We can also interpret coherence using the photon concept.
If the photons traversing the apparatus all have nearly the same
probability-density distributions in the interference region then
the apparatus is illuminated coherently, since all photons will
be contributing to the same fringe pattern. In practice, this
means that the photons must all have nearly the same momentum

and must come from nearly the same source point. This then
corresponds to the classical condition, since momentum includes
frequency and direction.

The above explanation is only a sketchy attempt to explain a topic
which is in detail very complicated. However, much the same
argument applies to interference and diffraction of all
elementary particles. From a particle viewpoint we are dealing
with varying probabilities of arrival at different points, and
the wave picture provides, in effect, a convenient way of
calculating these probabilities. Then, after calculating the
probabilities by using wave theory, we can call them intensities
if we are dealing with a large enough flux of particles.

PROBLEMS

3.1. Two beams of radio waves of frequency 3 MHz intersect
at an angle of 10°. What is the interference-fringe
spacing?

3.2. How many fringes are formed per millimetre if light
beams of wavelength 632.8 nm intersect at 5°?

3.3. Two beams interfere at an angle θ. If the complex
amplitudes are in the ratio 2:1, show that the intensity
in the fringe system has the form

$$E^2\{5 + 4 \cos\left(\frac{2\pi}{\lambda} y\sin\theta\right)\},$$

and plot this function.

3.4. Two glass plates are nearly in contact and make a small
angle θ with each other. Show that the fringes produced
by interference in the air film have a spacing equal to
$\lambda/2\theta$ if the light is incident normally.

3.5. A monochromatic source of wavelength 546 mm is 25 mm in
diameter and is placed 500 mm above an air film between

two glass plates. Show that the air film can be about
0.2 mm thick before the fringes begin to lose visibility.
(*Hint.* The range of angles of incidence is from 0° to
$\theta = \arctan 12.5/500$; the range of path differences is
from $2d$ to $2d\cos\theta$, and this should be less than $\lambda/4$.)

3.6. An air film is 100 μm thick and fringes are to be
formed in it from a polychromatic source of mean wavelength 550 nm. Approximately what wavelength range can
be used? (Choose a range from λ_1 to λ_2 such that N does
not vary by more than 1/4.)

3.7. Plot the functions sinc x and $\text{sinc}^2 x$ for values of x up
to the third zero. Tabulate the values of the subsidiary
maxima.

3.8. An aperture 5 mm in diameter diffracts light of wavelength
0.5 μm. How far away must a screen be placed to show the
far-field diffraction pattern?

3.9. A teaching laboratory has a 2 m long optical bench.
Suggest a suitable aperture size for demonstrating far-
field diffraction with a helium-neon laser.

3.10. Plot the amplitude and intensity distribution in the Airy
pattern. Find by numerical or graphical approximation the
intensity in the first bright ring, the radius of the first
dark ring, and the radius at which the intensity is half
the central maximum.

3.11. A beam from a ruby laser (694 nm wavelength) is to be
used in measuring variations in the distance of the moon by
timing its return from mirror systems arranged on the moon.
If the beam is expanded to 1 m diameter and collimated,
estimate its size at the moon. (Moon's distance
~ 3.8×10^5 km)

4. Polarization

EVERYDAY ASPECTS

Most of us have observed polarized light, through 'Polaroid' sunglasses. What we are seeing arises because light can have asymmetry about the direction of propagation; the appearances change when the glasses are rotated. Thus in the wave representation of light the disturbance cannot be along the direction of travel, as it is in sound waves. Similarly, television aerials also have directionality, indicating that they are sensitive to e.m. waves in which the disturbance is perpendicular to the direction of propagation.

In radio waves it is the electric field which is mainly of importance in detection, and thus the direction of the electric field is the polarization direction. We define the *plane of polarization* as a plane containing the direction of propagation and the electric field vector at any point in the radiation field. In the optical region the detection processes are associated almost entirely with the electric field (although in the photon picture this is not quite so easily formulated), so that the electric vector again defines the plane of polarization.†

Radio waves are polarized by emission from a transmitting aerial of the right shape, see *Electromagnetism* (OPS 1), but light from everyday sources is usually unpolarized. Thus in the model (p.13) of a light beam with randomly varying phases and amplitudes we have to add that the state of polarization, which we shall define below, is also changing rapidly and randomly

†The plane of polarization was originally chosen by arbitrary convention to be that containing, as we now know, the magnetic field vector, and this definition can be found in texts published up to about 50 years ago.

in time. This is explained by saying that the states of polarization of wave-trains emitted by different atoms have no fixed relationships. However, if we require polarized light, it is very easy to obtain from an unpolarized beam, by passing the beam through a polarizer.†

KINDS OF POLARIZED LIGHT

To find out what a polarizer does we must first describe polarized light. Suppose we have a beam of collimated light travelling in the z-direction, as in Fig. 4.1. To polarize it we put a polarizer in the beam with its polarizing direction parallel to the y-axis,‡ and we then have the beam polarized with the electric vector in the y-direction. If we follow this with a second polarizer with its axis at right-angles to that of the first then no light is transmitted. If the second polarizer

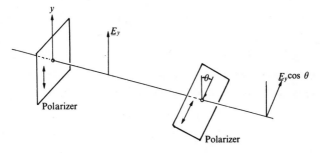

FIG. 4.1. The light transmitted by two polarizers at an angle θ.

†In the rest of this chapter we shall use 'polarizer' to mean a piece of 'Polaroid' or any other device, such as a Nicol prism, which can produce polarized light.

‡To find the polarizing direction, we use the polarizer as in sunglasses, i.e. we look through it at a horizontal smooth shiny surface and turn it so that the reflection from the surface is minimized. The polarizing direction of the polarizer is then vertical, as will be seen on p.67.

has its axis at some other angle θ we resolve the incident
electric field E_y into components $E_y\cos\theta$ and $E_y\sin\theta$ parallel
and perpendicular to the new direction, and only $E_y\cos\theta$ is
transmitted. Thus we should expect the transmitted light
intensity to vary as $\cos^2\theta$. Experimentally this is found to
be so, and this is confirmation of our model of a polarizer and
of polarized light.

The light produced by a polarizer as described above is
said to be plane-polarized, because the electric vector remains
parallel to one plane - that which contains the direction of
propagation and the polarizing direction of the polarizer. We
can propose other kinds of polarization, as follows. Light
plane-polarized in the y-direction has an electric field of the
form

$$\underline{E}_1(t,\underline{r}) = (0, E_y, 0), \tag{4.1}$$

where $E_y = E_2\cos(\omega t - kz)$ and the notation (A_x, A_y, A_x) denotes
the three components of the vector $\underline{A}$. These equations are
merely special forms of eqn (1.5) in which the real component
only is taken. We can now suppose added to this a coherent beam
travelling in the same direction but polarized at right-angles
and with different phase and amplitude,

$$\left.\begin{aligned} \underline{E}_2(t,\underline{r}) &= (E_x, 0, 0), \\ E_x &= E_1\cos(\omega t - kz + \varepsilon) \end{aligned}\right\} \tag{4.2}$$

(the experimental details of this addition are described on
p.69). To examine the result we take $z = 0$ and we suppose first
that there is no phase difference, i.e. $\varepsilon = 0$. Then at all times
the resultant electric field makes a constant angle $\arctan(E_2/E_1)$
with the x-axis, and so we have plane-polarized light, but with
the plane of polarization in the direction $\arctan(E_2/E_1)$. However
if there is a phase difference, the direction of the resultant
field will change with time; e.g. if $\varepsilon = \pi/2$

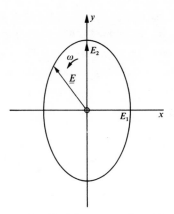

FIG. 4.2. The electric field vector in elliptically polarized
light. The vector rotates, tracing out the ellipse, with angular
frequency ω.

$$
\left.\begin{aligned}
E_x &= -E_1\sin\omega t, \\
E_y &= E_2\cos\omega t,
\end{aligned}\right\} \tag{4.3}
$$

and so the tip of a vector representing the electric field
traces out an ellipse(as in Fig. 4.2) with angular frequency ω.
We then have *elliptically polarized light*. In order to make the
distinction clear, we have to use 'plane-polarized' to describe
the effect of an ordinary polarizer. The most general effect of
adding two coherent plane-polarized beams with an arbitrary
fixed phase difference between them is to produce elliptic
polarization. If in eqn (4.3) $E_1 = E_2$, we have a special case –
circular polarization – since the ellipse given by the equation
becomes a circle.

In plane- , elliptic- , and circular-polarized light the
phase difference between two orthogonal components of the
electric field, resolved along any chosen axes, is constant in
time. Unpolarized light, sometimes called 'natural light', can
now be described, in a more general way than on p.13, as light
in which the state of polarization, in general elliptic, changes
rapidly and randomly in time. We can also have light beams
which are mixtures of unpolarized and polarized (plane, elliptic,

or circular) light.

In describing elliptic polarization a sign convention is necessary for the direction of rotation. The convention is that the rotation is clockwise looking towards the source for *right-hand* elliptic or circular polarization.

PRODUCTION OF POLARIZED LIGHT

The reflection factor, or ratio of reflected to incident light intensity, for a smooth interface between transparent media of different refractive indices can be calculated for e.m. waves (see e.g. Chapter 10 of *Electromagnetism*, OPS 1). Let the light have a angle of incidence θ_1 from a medium of index n_1, as in Fig. 4.3

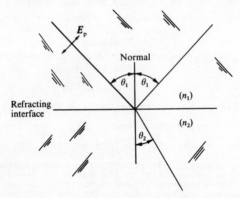

FIG. 4.3. Reflection and refraction of p-polarized light at a dielectric interface.

If the incident light is plane-polarized with the electric vector parallel to the plane of incidence, as in the figure, the reflection factor for light intensity is

$$R_p = \left(\frac{n_2/\cos\theta_2 - n_1/\cos\theta_1}{n_2/\cos\theta_2 + n_1/\cos\theta_1}\right)^2, \tag{4.4}$$

and for the other polarization, with electric vector perpendicular to the plane of incidence, the reflection factor is

$$R_s = \left(\frac{n_2\cos\theta_2 - n_1\cos\theta_1}{n_2\cos\theta_2 + n_1\cos\theta_1}\right)^2. \tag{4.5}$$

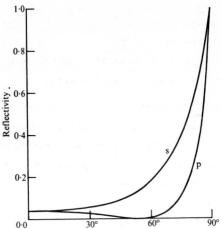

FIG. 4.4. Reflection factor of a glass surface for p- and s-polarizations as a function of angle of incidence.

The subscripts p and s, which are almost universally used, stand for *parallel* and *senkrecht* (German for parallel and perpendicular).

The ratios R_p and R_s are plotted in Fig. 4.4 for reflection at a glass-air interface, i.e. taking $n_1 = 1$, $n_2 = 1.5$. It can be seen that R_p is zero at an angle of incidence of about 57°, so that at this angle the reflected light will be completely plane-polarized perpendicular to the plane of incidence. It is easily shown that there is always an angle of incidence which gives complete polarization. This is called the *Brewster angle*, and from eqn (4.4) it is given by

$$\tan\theta_B = n_2/n_1. \tag{4.6}$$

This result accounts for the action of polarizing sunglasses in reducing glare from horizontal smooth surfaces. Even if the reflecting surface is not on a transparent medium there is still considerable polarization at oblique incidence, so that it is easy to find approximately the plane of polarization of a polarizer by looking through it at obliquely reflected light.

Light is often partly polarized by scattering as well as by reflection. Thus blue sky light is sunlight scattered by air

molecules, and it is strongly polarized for directions of
scatter at large angles to the incident beam. This is easily
seen through polarizing sunglasses.

An *anisotropic* medium is a medium in which the physical
properties vary with direction. All crystals except those with
cubic structure are anisotropic, and this is manifested as a
considerable complication in their optical properties. Consider,
for example, calcite (Iceland spar), which is crystalline
$CaCO_3$. We suppose that by some means it is possible to produce
a point source of light inside the crystal, as in Fig. 4.5.

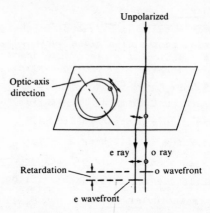

FIG. 4.5. Double refraction or birefringence in a crystal of
calcite ($CaCO_3$). The angle between the o and e rays is
exaggerated. The circles indicate that the electric field is
perpendicular to the plane of the diagram.

Then the effects observed can be explained by assuming that the
light does not propagate in simple spherical wavefronts, as in
an isotropic medium such as glass, but as in the figure. There
are two parts to the propagating disturbance, one a spherical
wavefront (the o disturbance, wavefront, or beam) and the other
an oblate ovaloid of revolution (the e disturbance). The two
systems propagate independently accordingly to Huygens' principle,
and they are polarized at right-angles as indicated. Thus if a
collimated beam is incident normally on the crystal, the o

disturbance will pass through undeviated, but if we propagate
the e disturbance as on p.23, by Huygens' principle we find it
is deviated as shown.† There is one direction in the crystal, the
direction of the *optic axis*, such that both disturbances travel
in the same direction and with the same velocity. From the
figure this is clearly the axis of revolution of the ovaloid.

In most crystals the effects are more complicated than as
described above, and it is necessary to consider propagation of
plane waves rather than a fictitious point source in the crystal.
In general, corresponding to any unpolarized plane wave incident
on the crystal from air, there are two plane-polarized waves
propagated inside the crystal in different directions and with
different speeds. Two beams plane-polarized at right-angles
and parallel in direction emerge from the crystal, as in Fig. 4.5.
The lateral displacement between the beams is used in some
polarizing devices, but a more important effect is a relative
retardation or optical path difference between the two emergent
wavefronts. The retardation is used to produce elliptic or
circular polarization, as in Fig. 4.6. Plane-polarized light,
with its plane of polarization at 45° to the optic axis of the
crystal, is resolved in the crystal into two orthogonal
components, with electric vectors respectively perpendicular and
parallel to the optic axis. The first becomes the o beam and
the second the e beam, and they propagate in the same direction
(since the optic axis is parallel to the surface) but with
different velocities. If the relative retardation when they
emerge is $\lambda/4$ or any odd multiple of $\lambda/4$ they combine, as on
p.65, to produce circular polarization, and the crystal plate is
said to be a quarter-wave plate. Other thicknesses (except those
giving $\lambda/2$ or a multiple of $\lambda/2$ retardation) would give elliptic-
polarization. It is customary to define a refractive index n_e

†The o and e rays were originally called the ordinary and
extraordinary rays on account of this behaviour.

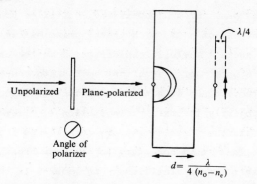

FIG. 4.6. A birefringent crystal as a quarter-wave retarder.
The optic axis is parallel to the crystal surface (normal to
the plane of the diagram as shown), and the thickness is chosen
to give $\lambda/4$ path difference throught the crystal between the o
and e rays. If the incident beam is polarized the two emergent
beams are coherent, and they produce circularly polarized light
for the arrangement shown.

for the e beam corresponding to the velocity of propagation,
although the e beam does not obey Snell's law on refraction at
the crystal surface. The retardation as an optical path length
is then $(n_e - n_o)d$, where d is the thickness of the crystal
plate.

Isotropic materials, e.g. glasses, become *double refracting*
or *birefringent*, like crystals, when under mechanical stress or
in static electric or magnetic fields. Materials such as
stretched plastic sheet (almost any kind) are found to be
birefringent on inspection between polarizers. These materials
contain long polymer molecules which acquire a partial common
alignment from the stretching, and thus they are similar to
crystals.. If a chromophore is added to the polymer it may
absorb one type of polarization and transmit the other; this
is the principle of the commonest kind of 'Polaroid', which is
a poly(vinyl alcohol)-iodine complex.

The effects of electric and magnetic fields in producing
birefringence have applications in modern optics,e.g. in
modulating the intensity in a light beam according to an

electrical signal. There is a second group of effects in which
a medium *rotates* the plane of polarization of incident plane-
polarized light. This effect is intrinsic in certain solutions
of molecules which form stereoisomers, i.e. the molecule and its
mirror image cannot be superimposed. A simple example is
lactic acid, $(CH_3)CH(OH).CO_2H$, in which the central carbon atom
is bonded to four different groups - methyl, hydrogen, hydroxyl,
and carboxyl, - so that this structure cannot match its mirror
image. Many organic compounds have this property, including
some sugars, and an important method, saccharimetry, of estimating
sugar concentration is based on it. Rotation of the plane of
polarization can be induced in almost all materials by a magnetic
field. For transparent materials this rotation is called the
Faraday effect. This effect has been used, for example, for.
estimating magnetic fields in space and for measuring very large
direct currents (by estimating the surrounding magnetic field).
Magnetic rotation also occurs on reflection from metal surfaces,
when it is called the *Kerr effect*. This effect is used in
studying the microstructure of magnetic alloys with a polarizing
microscope.

Mathematical formulations and a detailed e.m. theory treatment
of crystal optics and of electro- and magnetooptical effects
are given elsewhere (e.g. Born and Wolf 1965).

POLARIZATION AND INTERFERENCE

According to the discussion on p.40 we must include the state
of polarization in any precise discussion of coherence. If the
interference experiment as on p.63 is carried out with an
unpolarized source, we find interference fringes as expected.
However, if we polarize each interfering beam separately in
orthogonal directions there is no interference, and we cannot
produce interference by then rotating one of the planes of
polarization to agree with the other by means of, for example,

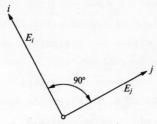

FIG. 4.7. Orthogonal plane-polarized components. They are
mutually incoherent if selected from an initially unpolarized
beam.

a half-wave plate (see Problem 4.6). We interpret this to mean
that unpolarized light is to be regarded as the sum of two plane-
polarized and *mutually incoherent* components which are orthogonal,
i.e. their planes of polarization are at right-angles to each
other. Thus in Fig. 4.7 we are looking along the unpolarized
beam, and we resolve the electric field into components along the
two orthogonal directions $\underline{i}$ and $\underline{j}$. Then the experiment shows
that the fluctuations in amplitude and phase which occur in $E_{\underline{i}}$
are entirely uncorrelated with those in $E_{\underline{j}}$. From this it follows
that a beam with an infinitely long coherence length, i.e. a
'perfect' laser beam, would be completely polarized, although not
necessarily plane-polarized.

The above result is sometimes put in the form 'beams polarized
at right-angles cannot interfere', but this is misleading. If
the initial beam in the interference experiment described on
p.40 had been polarized and then the split beams polarized as
above with polarizers at $\pm\ 45^{0}$ to the original direction of
polarization, we should not see interference fringes, but the
beams would be coherent and they would interfere. In fact this
is just the experiment described in Fig. 4.6, where the coherent
beams which are orthogonally polarized interfere to produce
elliptically polarized light. If a single component is selected
at the end by means of a polarizer parallel to the original
direction, then interference fringes can be seen.

To summarize this argument, orthogonally polarized beams do

not interfere if they are derived from two orthogonal components
of unpolarized light. They do interfere, producing in general
elliptic polarization, if they are derived from a single polarized
beam.

PROBLEMS

4.1.   A polarizer is placed in a beam of plane-polarized
       light of unit intensity, with its polarizing direction
       at an angle θ to the electric vector. Find (a) the
       modulus of the amplitude and (b) the intensity of the
       transmitted beam for $\theta = 10^{\circ}$, 45°, 88°.

4.2.   Elliptically polarized light has its axes along the x-
       and y- directions, and the relative field strengths are
       in the ratio 1:2. Calculate the relative intensity
       transmitted by a polarizer with its plane (a) parallel
       to the x-axis, (b) parallel to the y-axis, and (c) at
       35° to the x-axis.

4.3.   Prove eqn (4.6) by setting the numerator in eqn (4.4)
       equal to zero and eliminating θ_2 by means of Snell's
       law. Calculate the Brewster angle for materials of
       refractive index 1.5, 1.6, and 1.9.

4.4.   Prove that at Brewster incidence the reflected and
       refracted rays are mutually perpendicular.

4.5.   Calcite has $n_o = 1.659$, $n_e = 1.487$. Calculate the
       thickness of a quarter-wave plate for wavelength 589 nm.

4.6.   Show that the effect of a half-wave plate with its
       axis at an angle θ to the polarization direction of
       plane-polarized light is to rotate the plane of
       polarization through 2θ. (Resolve the incident field
       parallel and perpendicular to the plate axis.)

5. Image-forming instruments

Must get those old glasses of mine set right. Goerz lenses, six guineas.

James Joyce, 'Ulysses'

INSTRUMENT DESIGN

In the design of an optical system for forming images, any or all of three factors may be important: (1) light-gathering power, or the capacity to form a bright image; (2) magnification; (3) resolving power, or the capacity to form sharp images of small detail. There may, of course, be other factors which matter in particular cases, such as weight, size, method of transporting an image to a particular place, stability to temperature changes, etc., but we shall not discuss these. The main factors (1), (2), and (3) are often interrelated in particular systems, and it is thus convenient to discuss them in relation to actual instruments.

TELESCOPES

Fig. 5.1 shows the essentials of a refracting astronomical telescope. An objective lens, of diameter D and focal length f_1, forms a real image of stars and other astronomical objects at its focal plane. If the angle subtended between two of the stars is β then from Chapter 2 their images are separated by a distance βf_1. These real images are recorded by a photographic plate or other detecting system, or alternatively they may be viewed by the eyepiece. The eyepiece is a system of focal length f_2. It is used with the object, i.e. the real image of the star field, at its first focal plane. Thus it forms an image of the stars at infinity, so that they are viewed by the relaxed eye (probably equipped with glasses in the case of an

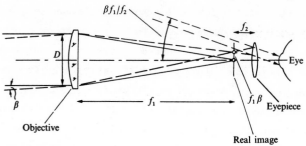

FIG. 5.1. A refracting astronomical telescope. The objective and eyepiece are shown conventionally as single-component lenses. Rays from two stars are shown, one on axis (full lines) and the other an angle β away from the axis (broken lines).

elderly astronomer). The pair of stars now subtends an angle $\beta f_1/f_2$, so that the angular magnification is f_1/f_2. Thus for physical detectors the magnification is determined by a scale factor, according to which an angle β between objects at infinity corresponds to a distance βf_1 in the image plane, whereas for visual observation we use the angular magnification f_1/f_2.

The light-gathering power depends on conflicting factors. Image-recording detectors, in the sense used on p.11, all have a certain minimum size of image which they can 'see'. Thus, suppose that by some means we produce an extremely small point image, say 250 nm across, on a photographic emulsion.† The image patch recorded by the emulsion will actually be much larger, owing to scattering and diffusion of light in the emulsion and other effects. Such an image produced by a negligibly small light patch is called a *point spread function*, and its size can be used to characterize the limiting performance of the detector. The concept is applicable to all detectors. In the human eye the point spread function projected back into the outside world corresponds to an angular subtense of about 0.0003 rad or a distance of 1 mm at 3 m; for a television camera it corresponds

†This is about the smallest point image which can be produced in the optical region, as will be seen on p.83.

to the scan-line size; for photographic emulsions it can range
from 500 nm (for special emulsions for holography or spectroscopy)
to about 0.02 mm for very high-speed panchromatic emulsions.

Returning now to the telescope, if the star image is smaller
than the point spread function of the emulsion, the light-
gathering power must be proportional to D^2, since it is simply
a question of the total of light flux collected. On the other
hand, if the image is larger than the point spread function, as
in the case of a planet or a nebula, the light-gathering power
is measured by the flux per unit area falling on the emulsion,
and then from p.33 it is proportional to D^2/f^2. In considering
light-gathering power for visual observation we have to ask
whether the pupil of the eye admits all the light collected by
the telescope. It can be seen by following rays through the
system that the eyepiece forms an image of the aperture of the
objective at a point to the right of the whole system where
pencils from a star away from the axis cross the axis. In this
context the aperture of the objective is the *entrance pupil*, and
this image of it is the *exit pupil*. Clearly the eye must have
its pupil roughly at the exit pupil of the telescope in order
to see all the field of view. Thus the light-gathering power
for visual purposes depends (1) on whether the exit pupil of the
instrument is larger or smaller than that of the eye, and (2) on
whether the star images are larger or smaller than the point
spread function of the eye. This argument is taken further
elsewhere (see e.g. Chapter 9 of Welford (1962)). The above
discussion shows that the photometry of optical systems can be
a complex topic.

A classical problem in astronomy is the resolution of double
stars or similar close objects. On the basis of geometrical
optics alone there need be no limit to resolving power; we
merely have to make a telescope with adequate magnification and
light-gathering power and with perfectly corrected aberrations.

However, according to physical optics, there is a limit. We regard the two stars to be resolved as point sources of equal brightness and not, of course, coherent with each other, since they are in fact separate thermal light sources. The telescope aperture, i.e. the rim of the objective, limits the size of plane wavefronts accepted from one of the stars and thus diffraction occurs at the aperture. The objective itself then brings the far-field diffraction pattern to a convenient place for viewing - the focal plane. In other words, the image of a point object according to physical optics is the far-field diffraction pattern of the aperture of the objective. We have already described this image on p.55 (eqn (3.16) and Fig. 3.12), and we know that it is a diffuse patch of light of which the angular size depends on the diameter of the diffracting aperture and on the wavelength of the light. From the present point of view we can call it the point spread function of the objective. Obviously the resolution of the optical system, leaving aside the effect of the detector, is determined by the size of this point spread function, since the image of the double star consists of two overlapping point spread functions. From the discussion on p.57 the angular half-width of a point spread function is of order of magnitude λ/D, and this therefore is roughly the limiting angular separation at which the two stars can just be seen separate through the telescope, i.e. it is the *angular resolution limit*. It is traditional to take the resolution limit as the separation at which the central intensity maximum of one point spread function falls on the first dark ring of the other (as in Fig. 5.2), and from Problem 3.10 (p.61) this gives for the angular resolution limit

$$\beta_{min.} = 1.22\lambda/D. \qquad (5.1)$$

In practice the difference between 1.22 and unity is rarely significant. Also in practice, the performance of ground-based

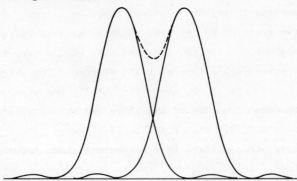

FIG. 5.2. Two star images of equal intensity which are
conventionally a just-resolvable distance apart. The angular
separation of the stars is 1.22λ/D, where D is the diameter of
the telescope objective. The sum of the two images is indicated
in broken line.

astronomical telescopes with objective diameters greater than
about 100 mm is limited by atmospheric turbulence, and the
very large diameters of the great telescopes are for light-
gathering power, not for resolution.

The above discussion shows how the factors listed on p.74 can
enter into any given problem. To summarize: (1) we must have
enough light-gathering power to record the required event in a
suitable time, e.g. up to several hours in astronomy, less than
a nanosecond in the study of pulsed lasers, or anything in
between; (2) there must be adequate magnification to ensure that
the detector can separate all the detail in the image which is
(3) resolved by the optical system. We must always make a clear
distinction between (2) and (3). For a telescope the resolution
depends only on the diameter of the objective and it is
independent of the focal length. On the other hand, again in the
case of the telescope, the scale of the star picture on the
photographic plate depends on the focal length but *not* on the
diameter of the objective. A similar distinction can be drawn
for most optical instruments.

Light-gathering power is the most important factor in modern

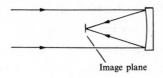

Image plane

FIG. 5.3. The Newtonian telescope. The photographic plate or other detector is placed at the image plane. (Newton's instrument was, of course, used visually with a small plane mirror to permit viewing from the side of the telescope tube: modern telescopes with a paraboloidal primary mirror are always called Newtonian.)

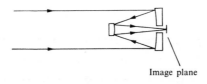

Image plane

FIG. 5.4. The Cassegrain telescope. The convex mirror, called the secondary, has a hyperboloidal shape if the primary is a paraboloid. The system has a relatively long focal ratio ($F/8$ to $F/11$) to match a spectroscopic system attached to the telescope.

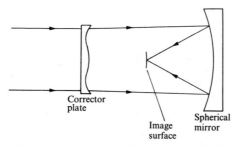

Corrector plate

Image surface

Spherical mirror

FIG. 5.5. The Schmidt camera. The essential feature is the corrector plate with a non-spherical surface. The main mirror is spherical, and the corrector plate provides the aberration correction which the main mirror would have if it were a paraboloid. However, since the corrector is at the centre of curvature of the mirror the same correction applies at all field angles, and thus the camera has a large, well-corrected field of view.

astronomy. For technical reasons concerned with chromatic
aberration and with the manufacture of optical glass all large
telescopes have mirror objectives. Figs. 5.3 - 5.5 show three
widely used types. The Newtonian telescope (Fig. 5.3) is simply
a large concave paraboloid of revolution. It is easily shown
that this brings incident parallel rays exactly to a point focus
at the (geometrical) focus of the generating parabola; the
detector, e.g. a photographic plate or a photoelectric image-
intensifier, is placed at the image plane. The Newtonian is the
system with greatest speed for direct image recording, since the
mirror can have a focal ratio of about $F/3$.† Sometimes a
spectrographic system has to be added to the telescope, and then
it is better to have a longer or larger focal ratio, say $F/8$ to
$F/11$, in order to match that of the spectrograph. The
Cassegrain system (Fig. 5.4) is then used. It is not usually a
separate telescope, but is obtained by adding an auxiliary convex
mirror to a Newtonian primary mirror; the larger focal ratio is
thus obtained without an increase in overall length of the system.

Both the Newtonian and Cassegrain telescopes have very small
angular fields of view, owing to off-axis aberrations. On a large
telescope the well-corrected field may be only about a minute of
arc. The Schmidt telescope (Fig. 5.5) can have an angular field
of a few degrees at about $F/2.5$ for apertures exceeding 1 m: it
is normally used photographically (and therefore usually called
the Schmidt *camera*) for rapid, large-scale surveys of the sky.

THE HUMAN EYE

We have to describe the eye both as an optical system and as
a detector, since it is used together with other optical systems.

†An objective is said to have a focal ratio or F-ratio of F/N if
its focal length is N times the diameter of its aperture or
entrance pupil. Thus a small focal ratio is photographically
very fast.

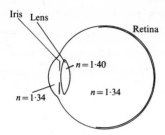

FIG. 5.6. The human eye. This is a very schematic and simiplified diagram. Most of the lens effect is due to the sharply curved front surface of the eye, the cornea. The iris is the aperture stop of the eye.

Fig. 5.6 is a very simplified diagram of the eye. Most of the refracting power is in the front surface of the cornea, and the main function of the lens is to vary the focus so as to be able to see clearly over a range of distances. Thus the eye is like a camera, forming an image of a distant scene on a light-sensitive surface, the retina. The retina is an array of light-sensitive cells which communicate with the brain *via* a complex network of interconnected nerve cells. The normal eye can adjust its focus or *accomodate* to form sharp images of objects at distances from infinity to about 250 mm.† The accomodation is not done as in a camera, by changing the distance between the lens and the detector, but by varying the lens curvatures by muscular control. The iris varies in diameter between about 2 mm and 8 mm, depending on the average brightness of the scene being viewed. The sensitivity of the visual channel from the retina to the brain also varies (*adaptation*), so that the eye can be used over a very wide range of light intensities, more than 8 orders of magnitude for diffusely illuminated scenes. A very faint flash consisting of only a few photons of green light striking a single receptor in the retina can be detected under

†But the range of accomodation decreases in old age. Many people have a different range, and they have to wear glasses in order to add or subtract refractive power to shift the range to the 'normal'.

suitable conditions, and at the other end of the scale a flux of 0.1 mW on a single receptor can be tolerated for about 0.1 s, so that on this basis the range of sensitivity of the eye is about 14 orders of magnitude. This range is obtained, as in most forms of sensory perception, by an approximately logarithmic response, i.e. a given *difference* in sensation corresponds to the same *ratio* between light signals at any part of the range. In many physical measurements a linear relation between the quantity measured and the indication is desired, but the great compression of range produced by a logarithmic response is often very useful.

As noted on p.75, the angular resolution of the eye is about 1 min of arc (0.0003 rad), depending on the conditions. This figure is used in determining the required magnification of an optical instrument. Thus if a telescope can *resolve* detail of 1 sec of arc we have to make the eyepiece magnify this detail enough to subtend 1 min to the eye (in practice 2 or 3 times more). The magnification is calculated as on p.75.

We must stress that the above description of the eye is very incomplete and lacking in detail. The responses of the eye to varying light levels, to fine detail, and to different wavelengths are very complicated, and are by no means fully understood. Our description is intended merely as a sketch to suggest how the eye is coupled to other optical systems.

THE MICROSCOPE

A microscope is essentially an elaborate magnifying glass. If we use a lens of focal length f to form an image at infinity of an object of size η (as in Fig. 5.7), the magnified object appears to subtend an angle η/f. On the other hand, if we use the unaided eye the object must be at least 250 mm away because we cannot accomodate nearer than this distance (p.81), so the object then appears to subtend an angle $\eta/250$. Thus the

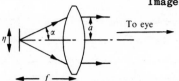

FIG. 5.7. A magnifying glass. The object is (approximately)
at the front focus, so that the image is seen by the relaxed
eye at infinity and it appears to subtend an angle η/f. In
practice the magnification is not very dependent on the exact
position of the object.

magnification is $250/f$, provided f is in millimetres. We must
also ensure that we can resolve the magnified detail. Thus from
p.77 the point spread function of the lens appears to the eye
to subtend an angle $\lambda/2a$, where $2a$ is the diameter of the
magnifier, or, turning this into a distance in the object space,
the point spread function is $\lambda f/2a$ across, and this must be
smaller than the detail to be resolved. Also a must be small
enough to ensure that all the rays get into the pupil of the eye.
The quantity a/f is roughly the semi-angle of the cone of rays
collected by the lens from the axial object point, and it must
be made as large as possible to resolve small detail, since the
size of the point spread function is proportional to f/a. In
fact it is found (Welford 1962) that it is the sine of this angle
which matters, and the accepted form of the resolution limit is

$$\eta_{\text{min.}} = 0.5\lambda/\sin\alpha. \qquad (5.2)$$

The precise value of the numerical factor depends in a complicated
way on the conditions of illumination of the object and on just
how we define 'resolution', but the value 0.5 is adequate for
most practical purposes.

Since the diameter $2a$ must be less than, say, 4mm in order to
match the pupil of the eye, it can be seen that we have to use
very short focal length lenses in order to get high resolving
power, and focal lengths of 2 mm or less are used with angles
α up to about 60°. The lenses have to have many components to
keep the aberrations small, and it is then found to be impossible

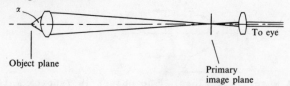

FIG. 5.8. Principle of the compound microscope. The
objective has a very small focal length so that the primary
image is highly magnified, and it has a large collecting angle
α for high resolution. The primary image may be recorded
directly on a physical detector, e.g. a photographic emulsion,
or it may be viewed through an eyepiece.

to get the eye close enough to the lens to see a reasonable
field of view. Thus the so-called compound microscope was
developed, as in Fig. 5.8. An objective with much the same
characteristics as we postulated for the magnifier in Fig. 5.7
is used to form an enlarged real image of the object. This real
image can be recorded photographically or by means of any other
physical detector, just as for the primary image in an
astronomical telescope; or alternatively we can form a virtual
image of it at infinity with an eyepiece and simply look at that,
as in conventional microscopy. Whatever mode of detection is
used we again note that the function of resolution depends on
the objective and its collecting angle, but not on the details
of the optical system which follows. Since the wavelength appears
in eqn (5.2) in the numerator we can also gain resolution by
going to shorter wavelengths. Little has been done with
ultraviolet-light microscopy, but great gains have been made in
electron microscopy. The de Broglie wavelength of electrons of
1 MeV energy is about 10^{-12} m (*Radiation and quantum physics*
(OPS.3)) and the collecting angle of electron lenses used as
microscope objectives is of the order of 10^{-3}, so that the
latest electron microscopes are approaching the resolution of
intermolecular distances. However, the collecting angle is
limited to that small value by aberrations which seem to be in
principle irreducible, whereas optical microscope objectives can

be made practically aberration-free for values of α up to 60°.

In eqn (5.2) λ in the numerator is the wavelength in the medium containing the object. Thus if the object is embedded in a medium of refractive index n then $\lambda = \lambda_0/n$, where λ_0 is the vacuum wavelength of the light. Eqn (5.2) then takes the form

$$\text{resolution limit } \eta_{\text{min.}} = 0.5\lambda_0/n\sin\alpha. \qquad (5.2a)$$

and we see that a gain in resolution is obtained by so embedding or immersing the object in a medium of high refractive index. This is the principle behind the oil-immersion microscope objective. The quantity $n\sin\alpha$ is called the *numerical aperture* (NA), and it is quoted on microscope objectives as a measure of resolving power.

IMAGES OF EXTENDED OBJECTS

Telescopes and microscopes have relatively small fields of view and the aberrations can be made small enough over these fields to justify taking the point spread function as the ideal Airy pattern (Fig. 3.12, p.56). This is not so with wide field systems such as objectives for film or television cameras, and then the point spread function formed with aberrations can be very different from the Airy pattern. Thus a 50 mm $F/1.8$ camera lens has the very large aperture $F/1.8$ so that short exposures can be used. If the lens had no aberrations its point spread function would be that corresponding to a convergence angle of about 20°, i.e. it would be less than 1μm in half-width, and this would be unnecessarily small for ordinary photographic emulsions. If an average is taken over a reasonable wavelength range, so that fine diffraction structure is smoothed out, the point spread function might appear as in Fig. 5.9. Aberrations will often cause the point spread function to be unsymmetrical.

We can obtain an expression for the point spread function from eqn (3.11b)(p.54), since in the present context this

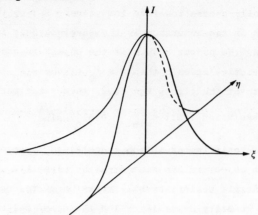

FIG. 5.9.   A point spread function. The height of the solid at
(ξ,η) is proportional to the light intensity in the point spread
function.

equation gives the complex amplitude in the point spread
function.   Eqn (3.11b) refers to a diffraction pattern in the
far-field, and so in order to bring it to the focal plane of the
camera lens we replace the angular coordinates u and v by ξ/f
and η/f, where ξ and η are linear coordinates in the focal plane
and f is the focal length. On p.54 we introduced the pupil
function $F(x,y)$, which defines the pupil area and allows us to
use the Fourier transform formalism.   The pupil function can be
used to define a pupil of any irregular shape, e.g. the
pentagonal iris diaphragm used in some cameras with automatic
exposure control, and in advanced work it is also made to include
the effect of aberrations.   Then if $I(\xi,\eta)$ is the light-intensity
distribution in the point spread function we have, from eqn (3.11b),

$$I(\xi,\eta) = \left| \frac{1}{A} \iint\limits_{-\infty}^{\infty} F(x,y)\exp\left\{\frac{\mathrm{i}2\pi}{\lambda f}(\xi x + \eta y)\right\}\mathrm{d}x\mathrm{d}y \right|^2. \qquad (5.3)$$

We cannot evaluate this expression until we know the form of
the pupil function $F(x,y)$, and often the integration has to be
done numerically.

We next consider how the image of an extended object such
as a bright disc or square is built up from individual point
spread functions. We assume the object to be incoherently
illuminated, i.e. the light from any one point of the object
cannot interfere with that from any other point, so that we
obtain the effect of overlapping point spread functions by adding
their *intensities*. This will be so if the object is self-
luminous, e.g. a hot filament or a gas discharge, or if it is
illuminated by a non-monochromatic source of large enough size.
Suppose then that the distribution of light intensity in the
object is given by $O(\xi,\eta)$. We use the same coordinates (ξ,η) as
in the image space to denote points which are object and image
in Gaussian approximation, by simply rescaling the object
coordinates according to the magnification of the optical system.
An element of the object, say $O(\xi',\eta')d\xi'd\eta'$, produces a point
spread function which redistributes the light from the element
over the image plane. Thus a point $P(\xi,\eta)$ receives light from
all the spread functions, and in particular from the spread
function at (ξ',η') it receives the contribution

$$I(\xi - \xi',\eta - \eta')O(\xi',\eta')d\xi'd\eta',$$

as in Fig. 5.10. The total effect at P is obtained by summing
over all the points (ξ',η') of the object; then provided the

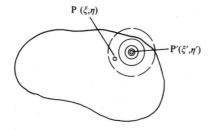

FIG. 5.10. The image of an extended object. The point spread
function centred at $P'(\xi',\eta')$ contributes some light intensity
at $P(\xi,\eta)$.

spread function has the same form at all these points we have

$$O'(\xi,\eta) = \iint I(\xi-\xi',\eta-\eta')O(\xi',\eta')d\xi'd\eta', \qquad (5.4)$$

and this is the light-intensity distribution in the image
$O'(\xi,\eta)$ of the object $O(\xi,\eta)$. Thus we see that the image is
obtained as the convolution of the object and the point spread
function (see Appendix). Convolution is the mathematical
representation of a physical process in which a sharply defined
input is spread to produce a blurred output. For example, in a
communication channel the output signal is the convolution of the
input signal and the *impulse response*. This latter is the
response of the channel to a delta-function input signal, and
so it corresponds to the point spread function in an optical
system.

We can formally write the object and image intensity
distributions $O(\xi,\eta)$ and $O'(\xi,\eta)$ and the point spread function
$I(\xi,\eta)$ as the inverse Fourier transforms of certain functions
$o(s,t)$, $o'(s,t)$, and $L(s,t)$,

$$O(\xi,\eta) = \iint o(s,t)\exp\{i2\pi(s\xi + t\eta)\}dsdt, \qquad (5.5)$$

$$O'(\xi,\eta) = \iint o'(s,t)\exp\{i2\pi(s\xi + t\eta)\}dsdt, \qquad (5.6)$$

$$I(\xi,\eta) = \iint L(s,t)\exp\{i2\pi(s\xi + t\eta)\}dsdt, \qquad (5.7)$$

and the convolution theorem (Appendix) tells us that

$$o'(s,t) = o(s,t)L(s,t). \qquad (5.8)$$

The physical significance of eqns (5.5) - (5.7) is that the
object and image are expressed as sums (i.e. integrals) of
sinusoidal components, s and t are *spatial frequencies*, e.g.
numbers of lines per millimetre, and eqn (5.5) means that the
periodic component in $O(\xi,\eta)$ with spatial frequency components
s and t in the x- and y-directions has the amplitude
$o(s,t)$ dsdt. Then according to eqn (5.8) the amplitude of the

periodic component in the image having the same pair of spatial
frequencies is obtained by multiplying the amplitude of the
object component by the factor $L(s,t)$, the Fourier transform
of the point spread function (eqn (5.7). This function
$L(s,t)$ is called the *optical transfer function*, usually
abbreviated to OTF, and its role in an optical system is
analogous to that of the transfer function of an electrical
channel such as an amplifier. The form of the OTF, i.e. its
numerical values, depends on the form of the point spread
function, but it can be shown that whatever the form of the OTF
a single-frequency object or sinusoidal grating forms a similar
(i.e. also sinusoidal) image, but with different contrast. The
reduction in contrast as a function of spatial frequency is
used as a measure of quality for optical systems which are to
be used for imaging extended objects in incoherent illumination.
Fig. 5.11 shows how images of sinusoidal objects of different
spatial frequencies are formed. The contrast is usually less
for higher spatial frequencies and there is no image contrast
at all, i.e. no modulation, for spatial frequencies above a
certain limit corresponding roughly to a grating with lines
spaced at the resolution limit (eqn (5.2)).

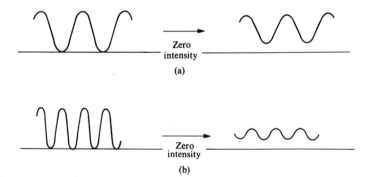

FIG. 5.11. Contrast transfer for sinusoidal objects. (a) and
(b) Objects of low and high spatial frequencies respectively,
showing greater reduction in contrast of the image, i.e. smaller
relative swing in intensity, for (b).

PROBLEMS

5.1. What is the theoretical angular resolution of
      telescopes with objective diameters of 100 mm, 1 m,
      and 5 m?

5.2. A telescope with a 50 mm aperture and 500 mm focal
      length objective is used for stellar photography.
      Estimate the size of a star image on the plate and
      calculate the plate scale in radians per millimetre.

5.3. The telescope in Problem 5.2 is to be used visually.
      What focal length would be required for the eyepiece
      in order to take advantage of the theoretical resolution
      of the objective?

5.4. Plot a graph of the light intensity across the centre
      line of two star images of equal intensities when the
      centres are separated by a distance equal to the radius
      of the first dark ring of the Airy pattern.

5.5. What is the resolution limit of a microscope objective
      of numerical aperture 0.65, and what overall
      magnification of the microscope is needed to take full
      advantage of this resolution?

5.6. Suggest a microscope NA and a magnification suitable
      for visual study of (a) red blood cells 7μm in diameter
      and (b) grains of silver halide 1μm in size in a
      photographic emulsion.

6. Interferometers and spectroscopes

Red rays are longest. Roygbiv Vance taught us: red, orange, yellow, green, blue, indigo, violet.

James Joyce,'Ulysses'

YOUNG'S EXPERIMENT; SPATIAL COHERENCE

Fig. 6.1 shows a simple way to produce interference effects. The two pinholes in the screen placed in the collimated beam act as secondary sources and produce divergent spherical wavefronts;

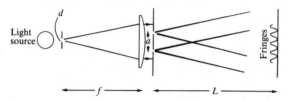

FIG. 6.1. A version of Thomas Young's interference experiment. The collimator is not essential. It is put in to make the phase relationships between the two pinholes clearer.

these interfere where they overlap, since they have come from the same original source, and they produce interference fringes as in the figure. A simplified version of this experiment was carried out by Thomas Young in 1804. The experiment led to the general acceptance of the theory that light is a wave phenomenon.

Young's experiment can be used to illustrate the concept of coherence between light beams. If the source in Fig. 6.1 is a helium-neon laser it is not essential to have a pinhole at the focus of the collimator and the fringes will have good contrast or visibility if the two secondary pinholes are equal in diameter. If we use a thermal source such as a sodium lamp we find experimentally that the collimator pinhole must be restricted in size for fringes of good contrast to be formed. We can see why this is by an argument similar to that on p.42. Let the source pinhole have diameter d, let the collimator focal length be f, and let the two secondary pinholes be a distance a apart. The spacing of the fringes formed by light from a given point in the source is $\lambda L/a$, where λ is the wavelength and L is the distance from the pinholes at which the fringes are observed. The phase difference at the screen between disturbances from source points on either side of the pinhole is $(2\pi/\lambda)ad/f$, so that the different source points form fringe systems displaced laterally by the fraction of a fringe $ad/\lambda f$, as in Fig. 6.2. Thus if $ad/\lambda f$ is of order of magnitude unity or greater there will be more or less uniform illumination, i.e. no fringes will be seen. In other words, we must have

$$ad/f \lesssim \lambda/4 \qquad (6.1)$$

for fringes of good contrast to be formed.

This equation has more than one interpretation. It tells us how small the source pinhole must be in order to get good fringe

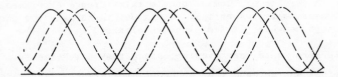

FIG. 6.2. Superposed fringes produced in Young's experiment by a source pinhole of finite size.

contrast: it must be smaller than $\lambda f/4a$, and it is then said to be a 'diffraction-limited pinhole'. Eqn (6.1) also tells us the distance between points on the screen within which the illumination is substantially coherent. This distance is about $\lambda f/4d$, i.e. this is the order of size of a 'coherence patch' from the source pinhole subtending the angle d/f. Thus the experiment illustrates the principle stated in Chapters 1 and 3 that there is coherence between light illuminating two different points if interference fringes of good contrast can be formed between beams of light from the two points. This is made quantitative in more advanced treatments by defining a degree of partial coherence according to the actual contrast of the fringes (Born and Wolf 1965). We see from eqn (6.1) that as the source gets bigger the coherence patch gets smaller.

We can use eqn (6.1) in yet another way. If we measure the size of the coherence patch from an inaccessible source we have a measure of the size of the source. This is the principle of the stellar interferometer invented by A.A.Michelson. Light from a star is collected by two mirrors of variable separation, as in Fig. 6.3, and the two beams are brought together and made to interfere. If the mirror separation at which the fringe contrast falls to zero is a the angular subtense of the diameter

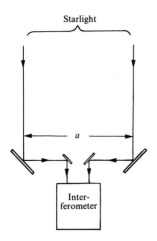

FIG. 6.3. Michelson's stellar interferometer. Beams a distance a apart on the ground but coming from the same star are made to interfere. The details of the interferometer do not matter in principle. The contrast of the interference fringes is a measure of the angular diameter of the star.

of the star is of order of magnitude λ/a. In Michelson's instrument the mirror separation a could take values up to about 6 m. The details of the arrangement by which the interference fringes are produced are in principle irrelevant to the

measurement: we merely have to make the beams interfere. The
same principle is applied in measuring the diameter of radio
stars, but the 'interference' is arranged by mixing the radio
signals collected by antennae at suitably variable spacings.
The wavelengths are in the centimetre range, and the
separations are of the order of a kilometre.

MICHELSON'S INTERFEROMETER; TEMPORAL COHERENCE

On p.42 we discussed briefly interference between beams
reflected at the two surfaces of an oil film on water or a
similar thin layer. The detailed theory of these effects is
complicated by multiple reflections to and fro in the film. The
principles can be seen more clearly in an apparently more
complicated apparatus, *Michelson's interferometer* (not to be
confused with his stellar interferometer), shown in outline in
Fig. 6.4. In principle, this is merely a device for studying
interference between coherent beams reflected from two parallel
or nearly parallel surfaces, but in order to avoid multiple
reflections the surfaces are not placed almost in contact, as
in the oil film. The surfaces are the two plane mirrors M_1

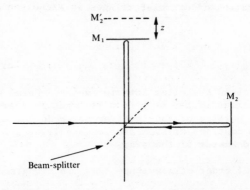

FIG. 6.4. Michelson's interferometer (not to be confused with
the *stellar* interferometer of Fig. 6.3), M_2' is the apparent
position of M_2 as seen in the beam-splitter.

and M_2. The beam-splitter, or semi-reflecting and semi-
transmitting mirror, is adjusted to make the image of M_2 appear
parallel to M_1 and at a distance z from it, as seen by a
detector (e.g. the eye) looking into the system as shown. We
consider interference between collimated beams from a source
at infinity having a finite angular subtense; this could be
arranged as in Fig. 6.5, where the source, e.g. a mercury lamp,

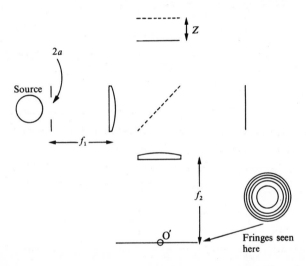

FIG. 6.5. Formation of circular fringes in Michelson's
interferometer.

is at the focus of a collimator of focal length f_1 and has a
diameter $2a$. The interference effects are to be observed at the
farfield, since this is where the individual source points are
imaged, and for this we use an objective of focal length f_2 and
observe at its focal plane.

A point of the source at a distance ρ from the axis of the
collimator produces a collimated beam inclined at an angle ρ/f_1
to the axis of the interferometer. This beam therefore meets
an effective plate of air of thickness z at an angle of incidence
ρ/f_1, and from p.42 the optical path difference between the

two beams reflected back from M_1 and the image of M_2 is

$$W = 2z\cos(\rho/f_1). \qquad (6.2)$$

At the centre of the far-field pattern this path difference has the value $2z$, and if this is an integral number of wavelengths the centre will be bright.† Going out from the centre, W decreases because of the cosine factor in eqn (6.2), and each time it decreases by one wavelength we reach another bright fringe. The fringes must be circular, since W depends only on the angle of incidence ρ/f_1, not on the azimuthal angle. It is easy to show that the radius of the Nth fringe is proportional to $N^{\frac{1}{2}}$.

Consider only the centre of the fringe system $0'$, and suppose the distance z to be varied steadily. Then, as the path difference $W=2z$ varies, $0'$ will be alternately bright and dark, and the intensity there will be proportional to

$$1 + \cos(4\pi z\nu/c), \qquad (6.3)$$

where ν is the frequency of the light. A detector at $0'$ would record this as a fringe pattern or *fringe function* in the variable z.

So far we have assumed the light to be monochromatic with frequency ν. If this is not so (e.g. we might be using a source with a broad spectrum line, such as a high-pressure mercury lamp, or perhaps a continuous source, such as a filament lamp) we suppose that the proportion of power in the light beam between frequencies ν and $\nu + d\nu$ is $G(\nu)d\nu$. Thus $G(\nu)$ is proportional to the light intensity seen through a prism or other spectroscopic system. The fringe function for the frequency band $d\nu$ is then

$$G(\nu)\{1 + \cos(4\pi z\nu/c)\}d\nu$$

and the total fringe function for the light of all frequencies

†Here, as elsewhere, we ignore complications due to phase-change effects on reflection.

added together is obtained by integrating with respect to ν,

$$g(z) = \int G(\nu)\{1 + \cos(4\pi z\nu/c)\}d\nu. \qquad (6.4)$$

The physical meaning of this equation is that we are adding together individual fringe functions of different fringe spacings. The fringe spacing in the z domain for frequency ν is $c/2\nu$. Thus these fringe systems start in phase at zero path difference ($z = 0$) and they gradually get out of phase as z increases, so that the contrast of the fringes falls. For nearly monochromatic light, i.e. a small range of frequencies, the contrast is good over a large path difference, and for 'white' light only a few fringes can be detected with measurable contrast. Thus the interferometer can be used to estimate the narrowness of a spectrum line. Michelson himself used it in this way in 1892 to show that the red cadmium line of wavelength 643.8 nm is very narrow and is therefore suitable for standardizing the metre in terms of wavelengths.

Returning to eqn (6.4) we see that the right-hand side is the sum of a constant (the integral over the spectrum of $G(\nu)$) and the cosine Fourier transform of $G(\nu)$. Thus if the fringe function $g(z)$ is recorded we can obtain the spectrum of the light by calculating the Fourier transform. This is the principle of *Fourier-transform spectroscopy*, which has become a standard technique in many fields within the last two decades. Michelson actually determined several spectrum line profiles in this way over 80 years ago.

The fall in contrast of the fringe function with increase of path difference 2z can be regarded also as decreasing coherence between the beams returning from the mirrors. Thus the path difference at which the contrast falls to some chosen value is a measure of the coherence length (p.15) and the corresponding coherence time is $2z/c$. These are therefore measures of the length of wave-train which is reasonably correlated with itself or which is approximately sinusoidal with the same frequency and amplitude along its length. Thus coherence length and spectral composition are two different aspects of the same physical phenomenon.

The Michelson interferometer of Figs 6.4 and 6.5 has many other applications (see e.g. Born and Wolf 1965), but that which we have described has the greatest importance in basic physics.

PRISMS AND GRATINGS AS DISPERSING ELEMENTS

The conventional spectroscope disperses the different wavelengths into different angular directions. The dispersing prism is still used essentially as in Isaac Newton's experiment

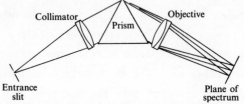

FIG. 6.6.   Principle of the prism spectrograph.

to produce a spectrum. Fig. 6.6 shows the principle of a *prism spectrograph*, used for photographing a spectrum. The deviation of a beam depends on the wavelength, since the refractive index of all materials varies with wavelength: since the deviation also depends on the angle of incidence of the light (as can be shown by a rather laborious application of Snell's law) it is necessary to collimate the light. The spectrum is formed at infinity with respect to the prism, i.e. in the far field, and therefore it is brought to a focus by the camera objective. Thus the complete system consists of a slit at the focus of the collimator to define the direction of the incident beam, the dispersing prism, the camera objective to focus the dispersed beams, and the photographic plate or

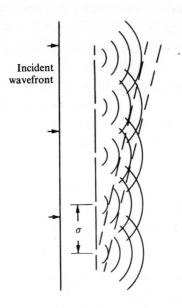

FIG. 6.7. First-order diffracted wavefronts as the envelopes of secondary waves from the slits or 'rulings' of a diffraction grating.

other image-recording detector. Fig. 6.6 shows only the principle of the simplest kind of spectrograph; there are many variations for different purposes.

The mode of action of a *diffraction grating* is not quite so obvious as that of a prism. Fig. 6.7 shows a grating consisting of narrow slits in an opaque screen with spacing σ between the slit centres. If a collimated beam of light is incident normally from the left each slit diffracts light as in Chapter 4, and, if the slits are narrow enough, the diffracted light spreads out over a range of angles, as indicated. The diffracted beams from each slit interfere, since they originated from the same collimated beam. In the far field all diffracted beams will be in phase in the direction of the original incident beam, but there can be other directions in the far field in which beams from neighbouring slits are one wavelength out of

phase with each other.   Such a diffracted beam is indicated in
the figure.

Fig. 6.8 shows how we can calculate the directions of this and
other diffracted beams: the ray and wavefronts are indicated
in inverted commas because they do not exist in the near field,

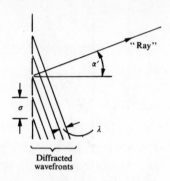

FIG. 6.8.   Calculation of the angle of diffraction.   The 'ray'
is the common normal to the 'wavefronts', and these are the
envelopes of the actual diffracted wavefronts, as in Fig. 6.7.

but the construction gives the direction of the diffracted ray
and wavefronts.   For the direction α' we must have

$$\sin\alpha' = \lambda/\sigma. \tag{6.5}$$

There may also be other diffracted beams with two,three, or more
wavelengths path difference between the waves from successive
slits, and for these we should have

$$\sin\alpha' = M\lambda/\sigma \quad (M \text{ an integer}). \tag{6.6}$$

Since in eqns (6.5) and (6.6) the angle of diffraction depends
on the wavelength, a collimated beam of white light incident on
the grating will be spread into a spectrum.   In fact there will
be several spectra corresponding to the different orders M, and
there will be an undispersed zero order, the light which travels
on undeviated.   This is indicated in Fig. 6.9.

Actual spectroscopic gratings are not made in Fig. 6.7.   The
distribution of light flux between the spectra of different

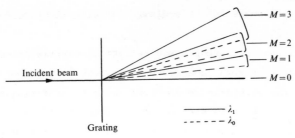

FIG. 6.9. Spectra of different orders formed by a grating.
λ_1 is greater than λ_0.

orders depends on the form of the individual slits or rulings;
any periodic structure acts as a diffraction grating (e.g. a
fabric of regular weave or the surface of a gramophone record),
and in a spectroscopic grating the form of the ruling, i.e. the
variation of transmission and phase across an individual slit
or line, is designed to throw as much light flux as possible
into a single diffracted order. Most gratings work in reflection
rather than in transmission, as in Fig. 6.10. By an extension
of the argument used above we can show that for angles of
incidence and diffraction α and α' the relation corresponding
to eqn (6.6) is

$$\sin\alpha + \sin\alpha' = M\lambda/\sigma. \qquad (6.7)$$

This equation gives the direction α' in which the Mth-order
diffracted beam of wavelength λ goes for an angle of incidence λ.

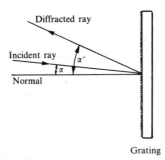

FIG.6.10. A plane reflection grating.

Obviously it reduces to the ordinary law of reflection for the zero-order beam.

Many grating spectroscopes use reflecting collimators and objectives rather than lenses because of the better aberration correction (*no* chromatic aberration) and greater wavelength range which can be used with mirrors. Fig. 6.11 shows one simple design of *monochromator*, i.e. a system with a second slit at the plane of the spectrum. Different wavelengths scan

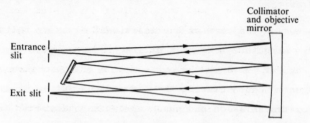

FIG. 6.11. A grating monochromator. The grating is rotated about an axis normal to the plane of the diagram to scan the spectrum across the exit slit.

across the exit slit when the grating is rotated. If the grating is formed on a concave surface the image-forming properties of the concave mirror are combined with the dispersion of the grating, and the spectrum is formed and focused by a single element, the *concave grating*, as in Fig. 6.12. If a complete spectrum is recorded photographically, as in the figure, the instrument becomes a *grating spectrograph*.

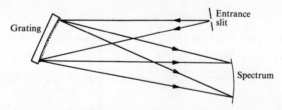

FIG. 6.12. The concave diffraction grating arranged as a spectrograph.

DISPERSION, RESOLUTION, AND LIGHT-GATHERING POWER OF PRISMS
AND GRATINGS.

The light-gathering power of spectroscopic systems depends
on the area of the entrance slit and on the angular subtense
of the collimator aperture in much the same way as for ordinary
image-forming systems, but it also depends on the dispersion,
i.e. the angular separation in the far field per unit wavelength
or frequency interval. The light-gathering power also depends
on the nature of the detector, i.e. whether this is an image-
recording system, such as a photographic emulsion in a spectrograph, or a total flux collector, such as a photomultiplier
at the exit slit of a monochromator. In the simplest case, a
monochromator with entrance and exit slits of equal width and
with a dispersing element of area A, the flux transmitted in the
wavelength interval $\delta\lambda$ is proportional to $\beta\delta\alpha A\delta\lambda$, where $\delta\alpha$ is
the angular subtense of either slit along the direction of
dispersion and β is the angular subtense of the height of either
slit. This can be written

$$\beta\delta\alpha . A \frac{d\lambda}{d\alpha'} . . \delta\alpha' , \qquad (6.8)$$

where $d\lambda/d\alpha'$ is the angular dispersion of the prism or grating.

We calculate the angular dispersion of a grating by
differentiating eqn (6.7) with respect to α',

$$\frac{d\lambda}{d\alpha'} = \frac{\sigma}{M}\cos\alpha' . \qquad (6.9)$$

Thus for small angles of diffraction the dispersion of a grating
is almost linear, since $\cos\alpha' \sim 1$ for small α', i.e. the wavelength found in the spectrum is directly proportional to the
angle of diffraction.

Fig. 6.13 shows the notation used for deriving the formula
for the dispersion of a prism. By differentiating Snell's law
for both faces and eliminating the internal angles of incidence

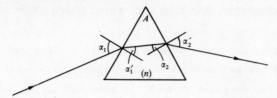

FIG. 6.13. Notation for calculating the dispersion of a prism.

it can be shown that

$$\frac{d\alpha_2'}{dn} = \frac{\sin A}{\cos\alpha_1'\cos\alpha_2'} \ ,$$

or

$$\frac{d\alpha_2'}{d\lambda} = \frac{\sin A}{\cos\alpha_1'\cos\alpha_2'} \cdot \frac{dn}{d\lambda} \ , \tag{6.10}$$

where $dn/d\lambda$ is the dispersion of the prism material. It can also be shown that in the symmetrical position, with $\alpha_1 = \alpha_2'$, the total angular deviation of the beam is a minimum, and for this *minimum deviation* position we have

$$\frac{d\alpha_2'}{d\lambda} = \frac{2\sin\frac{1}{2}A}{\cos\alpha_2'} \cdot \frac{dn}{d\lambda}. \tag{6.11}$$

For both the diffraction grating and the prism the final image is formed as the far-field diffraction pattern of an aperture which may be either the rectangular outline of the prism or grating itself or the aperture of a lens or mirror used to bring the far-field pattern to a focus. Thus an indefinitely narrow entrance slit illuminated with perfectly monochromatic light would still produce a spectrum line of a certain finite angular subtense. This, by the same reasoning as we used on p.77, must be of order of magnitude λ/D, where D is the width of the prism or grating aperture across the direction of dispersion. The criterion for the just-resolvable separation of two wavelengths λ and $\lambda + \delta\lambda$ is that their directions shall be

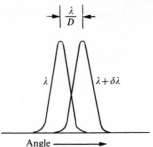

Fig. 6.14. The concept of spectroscopic resolving power.

separated by this angle, as in Fig. 6.14. The *resolving
power* of a dispersing element is conventionally defined as
the number

$$\lambda/\delta\lambda, \tag{6.12}$$

and we can find this for a grating by putting $d\alpha' = \lambda/D$ in
eqn (6.9). We obtain

$$\frac{\lambda}{\delta\lambda} = \frac{MD}{\sigma\cos\alpha'} ,$$

but, recalling that D is measured across the width of the beam
diffracted to the far field as in Fig. 6.15, we see that

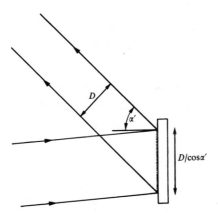

FIG. 6.15. Resolving power of a plane grating.

$D/\sigma\cos\alpha'$ is the number N of rulings on the grating. Thus we have

$$\lambda/\delta\lambda = MN, \qquad (6.13)$$

or, *the resolving power of a diffraction grating is the product of the number of rulings and the order of diffraction.*

Modern spectroscopic gratings for the visible region of the spectrum can be 50 - 300 mm wide and can have 300 - 1000 rulings per millimetre (but these are not absolute limits), so that a resolving power of order 3×10^5 is theoretically possible in the first-order spectrum. For comparison a spectrum line from a low-pressure gas discharge might have a width due to Doppler broadening of about 10^{-6} of its wavelength, so that its structure could be nearly resolved by such a grating.

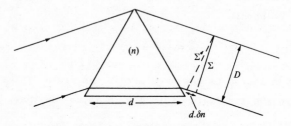

FIG. 6.16. Resolving power of a prism.

We can obtain the resolving power of a prism by similar reasoning to that used above, but it is easier to proceed from first principles. Fig. 6.16 shows a collimated beam of wavelength λ traversing a prism, with a plane wavefront Σ of the transmitted beam. If the wavelength is changed to $\lambda - \delta\lambda$ the wavefront will rotate to the position Σ' because of the increase in optical path length through the base of the prism; but this increase is $\delta n.d$, where d is the length of the path through the base and δn is the change in refractive index corresponding to the wavelength change $\delta\lambda$. In order to make the rotation equal to λ/D we have to make this change in optical path equal to one wavelength, i.e. we put $\delta n.d = \lambda$. Thus we find

$$\frac{\lambda}{\delta\lambda} = d \cdot \frac{dn}{d\lambda} \qquad (6.14)$$

This is the very simple formula for the spectroscopic resolving
power of a prism. Strictly d must be interpreted as the
difference between the extreme light paths across the beam,
since the edge of the prism would not be used.

MULTIPLE-BEAM INTERFERENCE

Between pp.91 and 98 we showed how the interference effects
between two beams, giving $\cos^2$ fringes, are used. The
diffraction grating (p.   ) works through the interference of
many beams, since it is only when the phases of all these
coincide (as in Fig. 6.7) that a maximum of light intensity is
found. In this context the grating is a *multiple-beam
interferometer*. If we illuminate a grating with collimated
monochromatic light we find maxima in the far field in directions
α', given by eqn (6.7), and we can plot these as in Fig. 6.17
with vertical broken lines corresponding to the order M. The
angular width of each maximum is, as on p.104, λ/D, where D is
the full width of the diffracted beam, and we have $D = N\sigma\cos\alpha'$.
Thus the angular width is

$$d\alpha' = \lambda/N\sigma\cos\alpha',$$

and the increment in $\sin\alpha'$ between successive orders is
$\Delta(\sin\alpha') = \lambda/\sigma$ from eqn (6.7). Thus

$$\frac{d(\sin\alpha')}{\Delta(\sin\alpha')} = \frac{1}{N} \qquad (6.15)$$

and we see that *each maximum is* $1/N$th *the width of the spacing
between maxima*, where as before N is the number of rulings in the
grating.

We said that the diffraction grating can be regarded as a

multiple-beam interferometer. From this point of view eqn (6.15) shows that the width of the interference maxima is found by dividing the interval between maxima by the number of interfering beams. This is a general property of multiple-beam interference effects, and it is in fact also true of two-beam interference, since two-beam fringes have equal light and dark spaces. Fig. 6.17 does not show in detail the distribution of

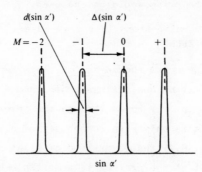

FIG. 6.17. Diffracted monochromatic light from a grating as an example of multiple-beam interference.

light intensity between the different maxima; for the detailed calculation, see, for example, Born and Wolf (1965).

Other forms of multiple-beam interference take place between parallel reflecting surfaces. On p.42 we discussed interference effects in thin films, but considered as an approximation only two beams - those first reflected from the upper and lower surfaces. These two beams would be beams 1 and 2 Fig. 6.18, but actually the light is multiply reflected as indicated by the broken lines, and strictly all the beams should be taken into account in calculating the interference effects.

In the case of the oil film or, for example, an anti-reflection coating of magnesium fluoride on glass, the effects of the succeeding reflections are negligible, since they are much fainter than the first two, but this is not so in other cases. Fig. 6.19 shows the *Fabry-Perot interferometer*. This

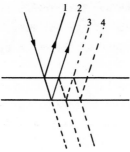

FIG. 6.18.   Multiple reflections in a thin layer.

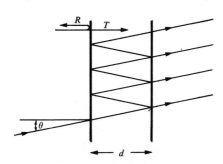

FIG. 6.19.   Principle of the Fabry-Perot interferometer.   The films, in practice supported on plates of fused silica, have high reflectivity and low transmission.   The angle of incidence is exaggerated to separate the rays, although in fact all the successive transmitted wavefronts overlap almost completely.

consists of two accurately plane and parallel surfaces with high reflectivity and low transmission (R and T respectively).   The spacing between the layers is d and collimated monochromatic light from a broad source falls on the first surface.†   We suppose the incident beam to be broad enough and the angle of incidence to be small enough to ensure that many multiply reflected beams emerge and interfere on the right-hand side.   The interference pattern in the far field can be brought to a focus by means of a

†The reflecting surfaces could be very thin silver layers supported on glass or fused silica.   They are usually dielectric multilayers, not silver, like the end mirrors of lasers, and typically we could have $R \sim 0.95$ and $T \sim 0.05$.

lens and, as for the Michelson interferometer, it must, by
symmetry, consist of concentric bright rings, each corresponding
to a certain angle of incidence θ at which all the beams are in
phase.

To find the form and spacing of the fringes we note that,
as on p.42, the optical path difference between successive
transmitted beams is $p = 2d\cos\theta$. Thus we can write down the
complex amplitudes of the successive transmitted beams as
follows, taking the origin of phase at the point of emergence of
the first beam,

$$T,$$

$$TR\exp(i2\pi p/\lambda),$$

$$TR^2\exp(i4\pi p/\lambda),$$

$$TR^3\exp(i6\pi p/\lambda),$$

$$\cdots\cdots\cdots.$$

Note that T and R refer to *intensities* but our present calculation is concerned with complex amplitudes, so that T gives the
amplitude transmission through two surfaces, as required, and
similarly for R. The total transmitted amplitude is the sum of
terms like these. To simply the calculation we assume the number
of terms is infinite, and we then merely have to sum a geometric
series with common ratio $R\exp(i2\pi p/\lambda)$. The result is

Transmitted complex amplitude =

$$= \frac{T}{1 - R\exp(i2\pi p/\lambda)}. \qquad (6.16)$$

To get the transmitted light intensity we take the squared modulus,
according to the rule on p.7 giving

$$\frac{T^2}{1 + R^2 - 2R\cos(2\pi p/\lambda)} = \frac{T^2}{(1 - R)^2 + 4R\sin^2(\pi p/\lambda)}$$

Finally, if we put $R + T = 1$, i.e. we neglect absorption in the reflecting layers, we have

$$\text{Fabry-Perot transmission } I(\theta) =$$
$$1/\{1 + \frac{4R}{(1-R)^2} \sin^2(\frac{2\pi}{\lambda}d\cos\theta)\}. \qquad (6.17)$$

It is easiest to see the general form of the multiple-beam
fringes by regarding $I(\theta)$ as a function of the phase difference
$\phi = (2\pi/\lambda)d\cos\theta$ which appears in eqn (6.17). For typical
values of R and T the quantity $4R/(1-R)^2$ is of order of
magnitude 1000. Thus the denominator in eqn (6.17) is very
large and the transmission is correspondingly small unless ϕ is
very close to a multiple of π. The transmission function is
therefore as in Fig. 6.10, i.e. the fringes have the
characteristics of multiple-beam interference as described for
the diffraction grating.

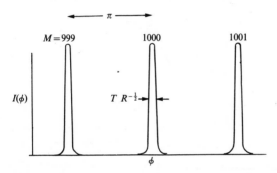

FIG. 6.20. The fringe shape for the Fabry-Perot interferometer
as a function of $\phi = (2\pi/\lambda)d\cos\theta$, half the phase difference
between successive beams. The exact fringe shape, given by
eqn (6.17), does not quite go to zero between the maxima. The
orders of interference indicated are notional.

The Fabry-Perot interferometer can be used as a spectroscopic
device and as an interferometer for the measurement of distance
or of phase. In either case we have to estimate the width of
the bright fringes and the spacing between fringes. A
convenient way of using the Fabry-Perot for spectroscopy is to
detect only the central fringe, as we described for the Michelson

interferometer, and to change the wavelength by varying the spacing d. It is then called a *scanning interferometer*. In this mode the maxima for wavelength λ must occur at spacings $d = M\lambda/2$, where M is an integer, the order of interference (but in the Fabry-Perot the order may be between 10^3 and 10^6, whereas in the diffraction grating it rarely exceeds 10). The orders are numbered typically in Fig. 6.20. Let the half-width of the fringes be given by 2ε as a fraction of the order - i.e. we suppose that, if we put $d = (M + \varepsilon)\lambda/2$ in eqn (6.17), the transmission falls to 0.5. Substituting in eqn (6.17) we have

$$\frac{1}{2} = 1/\{1 + \frac{4R}{(1 - R)^2} \sin^2(M\pi + \varepsilon\pi)\};$$

Removing $M\pi$ from the argument of the sine and putting $\sin \varepsilon\pi \sim \varepsilon\pi$, we find

$$2\varepsilon = \frac{1 - R}{\pi R^{\frac{1}{2}}} \quad .$$

This result is usually expressed in the form, fringe-spacing divided by fringe-width is

$$F = \frac{\pi R^{\frac{1}{2}}}{1 - R} \quad . \tag{6.18}$$

This quantity F is called the *finesse*.† We can use it to get an estimate of the resolving power of the Fabry-Perot interferometer. This is defined as for all spectroscopic devices as $\lambda/\delta\lambda$, where now $\delta\lambda$ is the measure in wavelength units of the quantity 2ε we found above. In the equation $d = M\lambda/2$ we now keep d constant and find the increment $\delta\lambda$ corresponding to an increment 2ε in M. We have

$$d = \frac{(M + 2\varepsilon)(\lambda - \delta\lambda)}{2},$$

†Eqn (6.15) gave the corresponding quantity for the diffraction grating, or rather its reciprocal.

or, to the first order in small quantities,

$$\lambda/\delta\lambda = M/2\varepsilon.$$

Substituting $1/F$ for 2ε we find

$$\lambda/\delta\lambda = MF, \qquad\qquad (6.19)$$

or, *the resolving power of the Fabry-Perot is the product of
the order of interference and the finesse.* By comparing eqn
(6.19) with eqn (6.13) it can be seen that the finesse plays a
role in the theory of the Fabry-Perot similar to the number N
of interfering beams in a diffraction grating. Thus F is
sometimes called the effective number of interfering beams.

In the Fabry-Perot a wavelength λ_1 may give a fringe maximum
of order M at the spacing d, while at the same spacing another
wavelength λ_2 may also have a maximum, but of order $M + 1$. Then
we have

$$M\lambda_1 = (M + 1)\lambda_2 = 2d.$$

The interval between these two wavelengths is the maximum length
of spectrum which can be studied without confusion between
spectra of different orders, and it is called the *free spectral
range.* There is a similar effect of overlapping orders with
the diffraction grating. Methods of avoiding confusion between
overlapping orders are described in books on spectroscopic
techniques.

SPECTROSCOPY IN GENERAL

In the preceding sections we have described only a few of
the large number of spectroscopic methods and instruments based
on apparently many different principles, but in fact there are
only a few underlying ideas, and it is mainly the technical
details concerned with adaptation for special purposes which
differ.

Spectroscopy is concerned with measuring the proportions of

different frequencies or wavelengths in a beam of polychromatic light. If we form a two-beam interference fringe system, e.g. with Young's apparatus (as in Fig. 6.1), the fringe spacing is proportional to the wavelength, and thus the intensity distribution in the fringe system formed by polychromatic light contains in an indirect or coded form the information we seek. The same is true for the Michelson interferometer - the fringe function is an encoded form of the spectrum. In both cases the decoding process is the same - taking the Fourier transform of the fringe function - but the Michelson interferometer gathers more light flux, and it is therefore preferable for most purposes.

The Michelson interferometer can also be regarded as a *multiplexer*. In telecommunications it is common practice to use a single line or channel to carry several messages simultaneously, e.g. by using the messages to modulate different carrier waves transmitted at the same time. Now if the mirror in the Michelson interferometer moves with velocity v the fringe function for monochromatic light oscillates at frequency $2v/\lambda$, as can be seen by putting $z = vt$ in eqn (6.3). Thus a wavelength λ is modulated at frequency $2v/\lambda$, and the interfero- meter is a multiplexing device which modulates each wavelength at a different frequency. The multiplexed signal is then decoded by taking its Fourier transform.

These ideas from communication theory have led to the development in the last 20 years of several new spectroscopic devices intended to increase light-gathering power or speed of operation. For example, we can have a grating monochromator with arrays of randomly spaced apertures in place of the normal entrance and exit slits, as in Fig. 6.21. Such a system would apparently have reduced resolving power corresponding to the greater wavelength range covered by the apertures, but if the prism or grating is oscillated at a certain frequency the central wavelength is strongly modulated at this frequency but

Direction of dispersion ⟶

Range of
oscillation

FIG. 6.21. Multiple slits, randomly spaced, of a modulation
spectrometer. The exit-slit array is similar. Wavelengths
throughout the whole spectrum are transmitted, but only the
wavelength for which the prism is set is strongly modulated.

neighbouring wavelengths are not.

Systems like the Michelson interferometer based on two-beam
fringes always produce an output which has to be decoded to give
the spectrum, and this is because the free spectral range of
two-beam fringes is zero. Thus in order to produce a spectrum
direct rather than encoded we have to use multiple-beam
interference, e.g. a diffraction grating or a Fabry-Perot
interferometer. In these and many other kinds of multiple-beam
spectroscopes the bright fringes are narrow enough to allow
fringes from many neighbouring wavelengths to be formed in
between them.

The dispersing prism is a special case outside these classes.
All the other systems we have mentioned rely on geometry alone
for their effects - diffraction at slits followed by interference,
or reflection at mirrors followed by interference. The prism
depends on a property of a material medium, namely dispersion,
whereas the other systems could operate in a vacuum with thin
films of conducting, i.e. reflecting material as mirrors, screens,
and beam-splitters.

In light scattered from a laser beam by, say, a colloidal
suspension or a turbulent gas stream there are fluctuations in
intensity which are, as explained in Chapter 1, interpreted as
a spread of wavelengths in the spectrum of the scattered light.
Let the intensity in the beam as a function of time be $I(t)$; we
can define the normalized autocorrelation function of the

intensity as

$$C(\tau) = \frac{1}{T} \int_{-\frac{1}{2}T}^{\frac{1}{2}T} I(t)I(t + \tau)dt$$

(see Appendix), where T is a time which is long compared to the fluctuations in question. From the autocorrelation theorem (Appendix) $C(\tau)$ is proportional to the Fourier transform of $\{G(\nu)\}^2$, the square of the spectrum of the light, so if we measure $C(\tau)$ we can obtain the spectrum. If the spectrum is very narrow, as it would be in scattered laser light, the delays for which $C(\tau)$ has to be measured can exceed 10^{-9}s, and then the autocorrelation can be measured directly by rapidly responding detectors which record individual photoelectrons. This technique is called *correlation spectroscopy*, and it has been developed in recent years as a method of spectroscopy suitable for very narrow spectral lines.

PROBLEMS

6.1.   In Young's interference experiment the source pinhole and the receiving screen are each 1 m from the two secondary pinholes, and these are 1 mm apart. (a) What is the fringe spacing for light of wavelength 546 nm? (b) Estimate the maximum diameter of the source pinhole for fringes of good contrast to be formed. (c) What would be the effect on the fringes if the two secondary pinholes were not of equal size?

6.2.   In an experiment with a stellar interferometer the fringes from the star Betelgeuse had zero visibility for a wavelength in the middle of the visible spectrum for a separation of the mirrors of 3 m. Estimate the angular subtense of Betelgeuse in arcseconds.

6.3.   For the Michelson interferometer of Fig. 6.5, find an expression for the radius of the Nth circular fringe from the centre of the far-field interference pattern, in terms of f_1, f_2, and λ.

6.4.   Calculate the form of the fringe function in a Michelson
        interferometer for a spectrum of rectangular profile.
        Plot a graph of this function for the case where the
        width of the spectrum in frequency units is 20 per cent
        of the mean frequency.

6.5.   The red line of cadmium (644 nm), as produced by a
        certain discharge tube, is found to have a coherence
        length of 200mm. Estimate the width of the line in
        wavelength and frequency units.

6.6.   A reflection diffraction grating has rulings with 1μm
        spacing. Draw a graph of the angle of diffraction as a
        function of wavelength for the first-order spectrum if
        the illuminating beam is at normal incidence.

6.7.   In the arrangement of Problem 6.6, if the shortest
        wavelength to be used is 200 nm, calculate the free
        spectral range, i.e. the wavelength range without over-
        lapping orders, for the first-order spectrum.

6.8.   A dispersing prism has an angle of 60°, and its
        refractive index is 1.762 for the mercury line of
        wavelength 546 nm. What is the angular deviation of
        this wavelength at the minimum deviation setting of the
        prism?

6.9.   The prism of Problem 6.8 has refractive index 1.791 for
        the mercury line of wavelength 436 nm. Estimate the
        spectroscopic resolving power of the prism in this
        region of the spectrum if it is equilateral with a 20 mm
        base.

6.10.   Draw a graph of the fringe function of a Fabry-Perot
        interferometer with $R=0.9$, $T=0.1$. Calculate (a) the
        finesse and (b) the minimum transmission.

6.11.   What is the spectroscopic resolving power of the
Fabry-Perot in Problem 6.10 if the spacing of the
plates is 5 mm? Calculate its free spectral range at
wavelength 500nm.

7. Laser light

Glass flashing. That's how that wise man what's his name with
the burning glass. Then the heather goes on fire.

James Joyce, 'Ulysses'

For the purposes of this book lasers are simply sources of very
intense, spectrally pure, and spatially coherent light.
Everything we describe could, in principle, be done with ordinary
thermal light passed through a monochromator of very high
resolving power and then through a diffraction-limited pinhole.
However, the light intensity would be so low that the experiments would in practice be impossible. Helium-neon lasers of
the kind now commoner than sodium lamps in many laboratories
produce about 1 mW of coherent monochromatic light of wavelength
632.8nm, but the brightest available thermal source, an ultra-
high pressure mercury lamp, would produce about 10 orders of
magnitude less light power of the same coherence and mono-
chromaticity.

COHERENT LIGHT SPECKLE

   If an optically rough or scattery surface is illuminated with
laser light a striking effect is seen: the surface is covered
with a fine network of bright and dark patches. This effect has
become known as *laser speckle,*although it can also be observed
with thermal light of sufficient coherence. The same effect is
seen in the far-field pattern from a rough surface, e.g. ground
glass, illuminated with a laser beam as in Fig. 7.1. The
explanation is that each point on the surface scatters a beam
which is coherent with the beams from all the other scattering
points, but there are random phase relationships between these
beams, so that in the farfield we see a superposition of

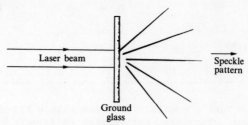

FIG. 7.1. Formation of a speckle pattern by a random diffuser. Speckle is formed at all distances from the diffuser, and the pattern has nearly the same statistical properties at all distances.

interference patterns between all pairs of scattering points. These interference patterns have random spatial frequencies, phases, directions, and contrasts, and the coherent sum of them all is the speckle pattern. The maximum spatial frequency in the pattern corresponds to interference between pairs of scattering points at opposite ends of a diameter of the ground glass.

The explanation of the speckle pattern seen by looking at the surface of the scatterer rather than at the farfield is slightly different. In looking at a certain point P on the surface we see the coherent sum of all the scattered beams within the radius of a resolution limit around P, and this coherent sum may again have a range of intensities, depending on the random phases of the scattered beams. This argument shows that the scale of detail seen in the speckle pattern on a scattering surface corresponds precisely to the resolution limit of the optical system used to view it.

Coherent light speckle obscures detail in the structure of the image of an object formed in coherent light and it is thus a great nuisance in image-forming systems which use coherent light. On the other hand, since it is an interference pattern, it may carry information about the statistics of the scattering surface, i.e. the height variations and their lateral scale, and also about movements of the surface; this is being turned to account in some recent applications.

HOLOGRAPHY

On p.40 we saw how two intersecting coherent collimated
beams produce straight and parallel interference fringes.
If the beams intersect at an angle θ the fringe spacing
perpendicular to the bisector of the angle θ is σ = λ/(2sin½θ).
If we record these fringes on a photographic plate or other
recording medium by placing it in the beam as in Fig. 7.2 and
then developing the plate, we have, in effect, a diffraction
grating (Chapter 6). We next set up the grating with one of

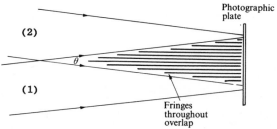

FIG. 7.2. Two collimated coherent beams form fringes throughout
the overlap region, with spacing ½λ/sin½θ. The plate records a
grating pattern with this spacing.

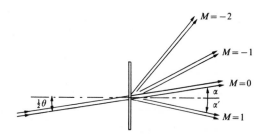

FIG. 7.3. Diffracted beams from the sinusoidal grating produced
as in Fig. 7.2. The first-order beam is the holographic
reconstruction.

the beams switched off, say beam 2, as in Fig. 7.3, and we find
several diffracted beams of different orders. The grating
equation (eqn 6.7) for a transmission grating can be written

$$\sin\alpha - \sin\alpha\acute{} = M\lambda/\sigma, \qquad\qquad (7.1)$$

and in this case we have $\alpha = \theta/2$. Then for the zero-order diffracted beam ($M = 0$) we have $\alpha\acute{} = \alpha$, i.e. an undeviated beam, and for $M = 1$ we find

$$\sin\alpha\acute{} = \sin\theta/2 - \lambda/\sigma;$$

but from the way we made the grating we have $\lambda/\sigma = 2\sin(\theta/2)$, so that for the first-order beam,

$$\alpha\acute{} = -\theta/2. \qquad\qquad (7.2)$$

This means that the first-order diffracted beam travels in the direction in which beam 2 of Fig. 7.2 was travelling. Thus we have 'reconstructed' beam 2 by illuminating the grating with beam 1.

Another way of looking at this process is to say that in photographing the fringe pattern we are attempting to record the complex amplitude of beam 2 at the plane of the photographic plate. The photograph does not record unambiguously everything about this complex amplitude distribution. Thus the fringe spacing tells us that the phase changes by 2π every fringe, but we do not know in which direction the phase is increasing. This ambiguity can be regarded as the cause of the appearance of the other diffracted beams of orders 0, -1,2,3, etc. On the other hand, if we had attempted to record beam 2 by placing the photographic plate in it without beam 1 to form fringes, we should have obtained merely a uniform blackening, i.e. a record of the intensity, and this could not contain any information about the direction from which beam 2 had come.

Beams 1 and 2 can be regarded as originating from point sources P_1 and P_2 at infinity. Then in the terminology of holography we formed a *hologram* of P_2 with P_1 as reference, and we then reconstructed P_2 with the same reference point and using the same wavelength light.

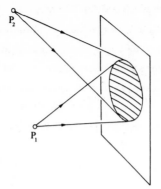

FIG. 7.4.   Formation of a hologram of a point P_2 with reference point.

A similar thing occurs if P_1 and P_2 are at finite distances, as in Fig. 7.4.   The fringes on the hologram are now curved, but it is again found that if the hologram is illuminated with P_1 alone then P_2 will be reconstructed (as in Fig. 7.5), and vice versa.

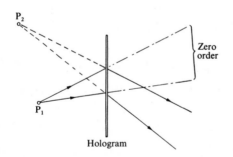

FIG. 7.5.   Reconstruction of the hologram of Fig. 7.4.   The first-order diffracted beam appears to come from the point P_2 behind the hologram.

Now suppose we have a reference point P_1 and an array of N object points P_2, P_3, . - . P_{N+1}, and we record the interference pattern as before.   The sets of fringes formed by interference between P_1 and each of the other points will all be formed, and, provided the recording medium has enough range, they will be superposed.   Then on reconstructing with P_1 the

image of the array of N points will be formed. This is the
principle of holography as invented by D. Gabor in 1948.

Ultimately the array $\{P_i\}$ becomes a continuous distribution, and
the hologram is a record of the interference pattern between the
reference beam and the complex amplitude scattered from this
continuous distribution. Let $E_0(x,y)$ be this complex amplitude
as a function of coordinates (x,y) in the plane of the hologram
plate, and let $E_r(x,y)$ be the complex amplitude due to the
reference beam. The light intensity in the interference pattern
is $|E_0 + E_r|^2$, and to a reasonable approximation the complex-
amplitude transmission T_a of the developed hologram is a linear
function of this,

$$T_a = k_0 - k_1|E_0 + E_r|^2,$$

where k_0 and k_1 are constants. In the reconstruction process
the complex amplitude transmitted by the hologram is $T_a E_r$, and
on multiplying out the squared modulus we see that this is

$$T_a E_r = k_0 E_r - k_1 E_r E_0 E_0^* - k_1 E_r^2 E_0^* -$$
$$-k_1|E_r|^2 E_0 - k_1 E_r|E_r|^2. \qquad (7.3)$$

The intensity $|E_1|^2$ of the reference beam is roughly constant
over the hologram, since it comes from a point source at some
distance. Thus the fourth term in eqn (7.3) is equal to a
constant multiplied by the complex amplitude $E_0(x,y)$ at the
hologram due to the original object, and this term therefore
accounts for the reconstructed image of the object. There are,
however, four other terms in eqn (7.3) to be accounted for.
It is easier to see the significance of these by assuming that
the reference beam is collimated and that it meets the hologram
plate at an angle of incidence θ, as in Fig. 7.6. Then the
complex amplitude in the reconstructing beam, assumed to be the
same as in the reference beam, can be taken as $\exp\{i(2\pi/\lambda)y\sin\theta\}$.

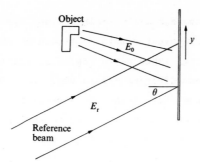

FIG. 7.6.   Forming a hologram of an extended object with an oblique reference beam.

Substituting this value in eqn (7.3) we obtain for the transmitted amplitude on reconstruction, after regrouping the terms,

$$T_a E_r = \{k_0 - k_1(1 + |E_0|^2)\}\exp\{i(2\pi/\lambda)y\sin\theta\} -$$

$$- k_1 E_0^*\exp\{2i(2\pi/\lambda)y\sin\theta\} - k_1 E_0. \qquad (7.4)$$

Of these terms the first represents a wave travelling in the same direction as the reconstructing beam but with different intensity; it corresponds to the zero-order diffracted beam in Fig. 7.3 and we show it as beam 1 in Fig. 7.7.   The second term,

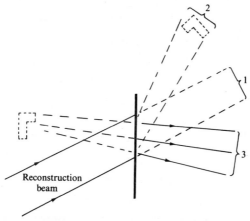

FIG. 7.7.   Reconstruction of the hologram of Fig. 7.6.   Beam 2, produced by the second term in eqn (7.4), produces another image of the object.   This image is real but usually very distorted. Beams 1, 2, and 3 correspond to the three terms of eqn (7.4).

on account of the exponent $2i(2\pi/\lambda)\sin\theta$, travels in a general
direction θ' such that $\sin\theta' = 2\sin\theta$. It can be shown to
produce a spurious real image as in Fig. 7.7, and it corresponds
to the diffracted beam of order - 1 in Fig. 7.3. The last term
produces the correct virtual-image reconstruction of the object.
The detailed properties of the photographic emulsion or other
recording medium control the relative intensities of these various
images.

Holographic images have the remarkable property of being
three-dimensional to the extent that the method of illumination
of the object and the angular range of viewing through the
hologram permit. This can be seen from Fig. 7.8, which shows a
typical arrangement of apparatus for holography.

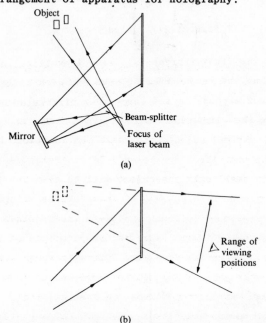

FIG. 7.8.   (a) A typical optical arrangement for taking a
hologram.   The laser beam is focused down by means of a
microscope objective and allowed to diverge.   (b) Reconstruction,
showing that the image can be seen from a range of angles to
give stereoscopy.

HOLOGRAM INTERFEROMETRY

The holographic image is a reconstruction of the complex amplitude of the light scattered from the object with a particular geometry of illumination. Suppose that a hologram of an object is taken but that the object is left in position at the reconstruction stage. If the hologram plate is replaced exactly in its original position after development it will produce a virtual image exactly coinciding with the object. In practice there will generally be a slight displacement between them, as in Fig. 7.9. The light coming through the hologram scattered from

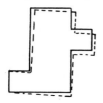

FIG. 7.9. A reconstructed image (broken line) superimposed on on the slightly displaced object. The displacement appears as a fringe pattern in hologram interferometry.

the actual object is coherent with the light from the reconstructed, virtual object, and there can therefore be interference between these two beams. If the relative displacement is small only one object will be seen but its surface will appear to be covered by interference fringes which indicate the relative displacement between the original object and its reconstructed image. This is the principle of hologram interferometry. It can be done with optically rough surfaces, in fact it is best done with rough surfaces, since then the illumination and viewing conditions are less critical. Classical interferometry, on the other hand, can only be done with smooth, mirror-like surfaces.

Hologram interferometry is used in engineering and metrology for determining displacements and strains of surfaces in a variety of applications. The mode of formation of the fringes

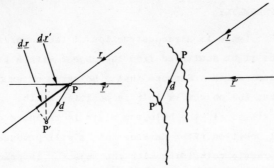

FIG. 7.10. The optical path difference mapped by the fringes
in hologram interferometry. The local displacement vector is
$\underline{d}$ (=PP´), and the local illumination and viewing directions are
specified by the unit vectors $\underline{r}$ and $\underline{r}´$.

is intrinsically more complicated than in classical interfero-

metry, as can be seen from Fig. 7.10. Each of the two surfaces,

that of the original object and its reconstructed image, appears

to the observer covered by a speckle pattern (p.119), and this

is itself an interference pattern. The 'fringes' seen in

hologram interferometry are due to interference between these

two speckle patterns. The speckle patterns themselves are

congruent but displaced, and therefore the displacement measured

by the fringes is that corresponding to the displacement of a

definite point, e.g. P, to P' in Fig. 7.10. This displacement

is in general not normal to the surface, and it may even be in

the plane of the surface, but the fringes will still map it as

an optical path difference.

   Let $\underline{r}$ and $\underline{r}´$ be unit vectors along the directions of

illumination and viewing near P and let $\underline{d}$ be the vector

displacement from P to P'. Then we see from the figure that the

total path difference indicated by the fringe system near P is

$$W = \underline{d} \cdot (\underline{r} + \underline{r}´), \qquad (7.5)$$

which is a rather complicated function of the directions $\underline{r}$ and

$\underline{r}´$ and the displacement. For example, we can measure a

displacement of a surface in its own plane, as in Fig. 7.11, if

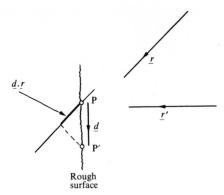

FIG. 7.11. Measurement of in-plane displacement by hologram interferometry. In the arrangement shown the surface is viewed normally, although this is not essential. The fringes map the quantity $\underline{d} \cdot \underline{r}$, which, as can be seen, is non-zero for an in-plane displacement.

the illuminating and viewing directions are suitably chosen, whereas such a measurement would give a null result in classical interferometry with smooth surfaces.

Hologram interferometry can be done in different ways. A variation of the method described above is to take two holograms on the same plate, one before and one after the displacement. The object is then removed and fringes are formed between the two reconstructed images. In another variation a hologram is taken of a vibrating surface with a time exposure lasting for many periods of the vibration. The reconstructed image carries fringes showing the form and amplitude of the mode of vibration.

HOLOGRAPHIC DIFFRACTION GRATINGS

We saw on p.121 that the photographed image of the fringes formed between two collimated beams forms a diffraction grating. This process can be carried out with a recording medium such as a photo-resist which forms a relief structure indicating the fringe distribution (as in Fig. 7.12), and after aluminizing a reflecting diffraction grating is formed. Techniques are

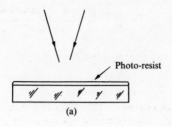

(a)

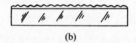

(b)

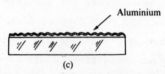

(c)

FIG. 7.12. Making a holographic diffraction grating. (a) A glass blank coated with photo-resist is illuminated with crossing laser beams to form the fringes. (b) The photo-resist is developed to give a contoured surface. (c) An aluminium reflecting coating is applied.

available for shaping the groove profile to produce a blaze, i.e. to direct most of the light into a single order of diffraction. Also the method can be applied to gratings on curved surfaces, and by careful choice of geometry better image formation can be obtained than in conventionally ruled gratings.

Hologram interferometry and the manufacture of diffraction gratings are probably the most important practical applications of holography at present.

SPATIAL FILTERING

We saw in Chapter 3 that the complex-amplitude distribution in the far-field diffraction pattern of an aperture with complex amplitude variations in it is the Fourier transform of these

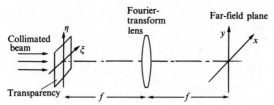

FIG. 7.13. Optical transforms. The complex amplitude in the farfield is the Fourier transform of the complex amplitude at the object transparency.

complex amplitude variations to a suitably chosen scale. This could be realized experimentally as in Fig. 7.13. A transparency placed in an aperture at the front focal plane of a lens is illuminated with collimated coherent light. The far-field diffraction pattern of the aperture and transparency is formed at the other focal plane, and from Chapter 3 the complex amplitude in this plane is

$$E(x,y) = \iint_{-\infty}^{\infty} T_a(\xi,\eta) \exp -\left\{ \frac{2\pi i}{\lambda f}(x\xi + y\eta) \right\} d\xi d\eta,$$

or, putting

$$x/\lambda f = s, \quad y/\lambda f = t,$$

$$E(s,t) = \iint_{-\infty}^{\infty} T_a(\xi,\eta) \exp -2\pi i(\xi s + \eta t) d\xi d\eta. \qquad (7.6)$$

Here s and t are spatial frequency components of the complex-amplitude transmission $T_a(\xi,\eta)$, just as we defined spatial frequency components of an intensity distribution on p.88. Thus the complex amplitude at (x,y) in the Fourier plane is proportional to the amount of complex amplitude with spatial frequency components $(x/\lambda f, y/\lambda f)$ in the original transparency. To take a simple example, suppose the transparency is a sinusoidal amplitude grating of complex-amplitude transmission

$$T_a = 1 + \cos 2\pi s_0 \xi;.$$

If we write this in the form

$$T_a = 1 + \tfrac{1}{2}\exp(2\pi i s_0 \xi) + \tfrac{1}{2}\exp(-2\pi i s_0 \xi),$$

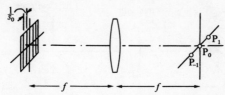

FIG. 7.14. The transform of a grating with amplitude transmission of sinusoidal form. If the grating spatial frequency
is s_0 the +1 and -1 orders are $\lambda f s_0$ from the zero order in the
transform plane.

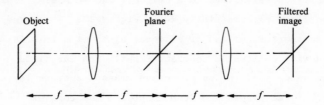

FIG. 7.15. A spatial filtering apparatus. The filters are
placed in the Fourier or far-field plane, and the filtered image
appears at the right.

we see (Appendix) that the Fourier transform consists of delta
functions of magnitudes 1, $\frac{1}{2}$, and $\frac{1}{2}$ at the origin and at
$(\pm s_0, 0)$. Thus in the Fourier plane we should see bright spots
P_0, P_{-1}, P_1 at the centre and at $(\pm\lambda f s_0, 0)$, as in Fig. 7.14.
From the point of view of diffraction-grating theory these are
merely the spectra of order zero and ± 1, and it would be
possible to develop the theory of diffraction gratings along
these lines.

Now suppose we add to the optical system of Fig. 7.13 another
lens of focal length f, as in Fig. 7.15. By elementary
geometrical optics we expect to find, at the second focal plane
of this lens, an image of the original object, since this object
is imaged at infinity in the intermediate space; this is found
to be broadly correct. However, if we were to obstruct part of
the Fourier plane we should be removing some of the spatial
frequency components, and the image would appear changed
accordingly. In the above example, the intensity distribution

in the object transparency is $|T_a|^2$, i.e.

$$I_0(\xi) = 1 + 2\cos 2\pi s_0 \xi + \cos^2 2\pi s_0 \xi$$

$$= \frac{3}{2} + 2\cos 2\pi s_0 \xi + \tfrac{1}{2}\cos 4\pi s_0 \xi, \qquad (7.8)$$

and the image would be the same. Now suppose that we put a mask
in the Fourier plane which removes the zero-order component,
i.e. the bright spot P_0. The complex amplitude in the image will
then be

$$T_a^{\prime} = \cos 2\pi s_0 \xi,$$

and the intensity will be the squared modulus of this,

$$I_a^{\prime}(\xi) = \tfrac{1}{2} + \tfrac{1}{2}\cos 4\pi s_0 \xi. \qquad (7.9)$$

Comparing eqns (7.8) and (7.9) we see that there has been a
complete change in the appearance of the image. Remembering
that what we see or detect is intensity, not complex amplitude,
eqn (7.8) represents a periodic structure with basic spatial
frequency s_0 (but also with a harmonic of frequency $2s_0$),
whereas in eqn (7.9) the basic frequency is $2s_0$.

Ernst Abbe explained these effects physically 100 years ago
in developing the theory of the resolving power of the microscope.
Abbe suggested that the image formed in the final image plane
could be regarded as an interference pattern between sources in
the Fourier plane. Thus in our example the three spots in the
Fourier plane in Fig. 7.14 produce intersecting coherent plane
waves in the final image space, and the image is the interference
pattern between these plane waves. This pattern will have a
basic periodicity corresponding to the angle between the waves
from P_{-1} and P_0 (or P_0 and P_1). If P_0 is removed the
interference pattern changes and its basic frequency is doubled,
since it corresponds to the angle between the waves from P_{-1}
and P_1.

Abbe also pointed out that if P_{-1} and P_1 are removed there will be no interference, merely uniform illumination in the image plane from a single point P_0. If we regard the two lenses as forming an imaging system, this shows why a certain minimum aperture is necessary to resolve a given spatial frequency. Specifically, the angular subtense of the semi-aperture must be at least P_0P_1/f; i.e. the collecting semi-angle α of the first lens must satisfy

$$\sin\alpha \geq \lambda s_0.$$

Thus the minimum resolvable separation $1/s_0$ is

$$1/s_0 = \lambda/\sin\alpha \qquad (7.10)$$

for spatial frequencies in complex amplitude.† This may be compared with eqn (5.2), which refers to incoherent illumination. The functional dependence on wavelength and aperture is the same, but the proportionality constant is different.

Returning to spatial filtering, it is possible to put at the Fourier plane a filter to change either the phase or the amplitude of certain frequency components in any desired way. As a simple example, the dot structure in half-tone pictures may be filtered out in this way. If a transparency of, for example, an aerial photograph has prominent features running parallel to a certain direction, these may be filtered out by means of an opaque strip across the Fourier plane at right-angles to the direction. This device may make it easier to see other features.

The re-imaging stage of Fig. 7.15 can be regarded as taking the inverse Fourier transform of the complex amplitude distribution in the Fourier plane. Thus from eqn (7.6) the final image is the original object,

†Note, however, that because of the process of taking the squared modulus, a given spatial frequency in complex amplitude may be doubled in intensity.

$$T_a\,'(\xi,\eta) = \iint\limits_{-\infty}^{\infty} E(s,t)\exp\{2\pi i(s\xi + t\eta)\}dsdt. \qquad (7.11)$$

If we differentiate this relationship with respect to ξ we obtain

$$\frac{\partial T_a(\xi,\eta)}{\partial\xi} = \iint\limits_{-\infty}^{\infty} 2\pi i s E(s,t)\exp\{2\pi i(s\xi + t\eta)\}dsdt, \qquad (7.12)$$

so that the derivative of the original is obtained by putting a filter in the Fourier plane with complex-amplitude transmission $2\pi i s$ or, changing back to the actual coordinate in the Fourier plane, $2\pi i x/\lambda f$. This is a linear variation of amplitude transmission across the x-direction.

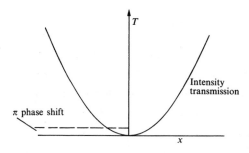

FIG. 7.16. A differentiating filter.

The factor i is not important, since it implies a constant phase shift over the whole plane, but since x changes sign on crossing the axis this implies a phase shift of π. Thus the filter would be as in Fig. 7.16. with a quadratic intensity transmission across the x-direction and a $\lambda/2$ film across one half. Many other devices of this kind are possible for analogue computing in the Fourier domain.

PROBLEMS

7.1.   A piece of ground glass of diameter D is illuminated uniformly by a laser beam, and a speckle pattern is formed on a screen at a distance L. Show that the smallest detail in the pattern is of order of size $\lambda L/D$. If the screen is viewed from 250 mm and if L is 2 m, how large must the ground glass be to make the smallest detail unobservable to the eye?

7.2.   In a holography experiment the reference source and the object are both 1 m from the hologram plate, and they are 100 mm apart. Estimate the scale of detail in the hologram fringes if the wavelength is 632.8 nm.

7.3.   A collimated reference beam is used to form a hologram of an illuminated pinhole. Sketch the arrangement and discuss the form of the hologram fringes.

7.4.   In a hologram interference experiment with a helium-neon laser the surface under test is illuminated and viewed at normal incidence. If the displacement to be determined is 2500 nm, how many fringes will it be represented by (a) if the displacement is normal to the surface; (b) if it is at 45° to the normal; and (c) if it is in the plane of the surface? In case (c) suggest a way of improving the sensitivity of the technique.

7.5.   A plane diffraction grating is to be produced holographically using light of wavelength 632.8 nm. Sketch the arrangement to be used, and calculate the required angles if the grating is to have 1000 rulings per millimetre. What is the closest grating spacing which could be made in this way?

Appendix: the Fourier transform and some of its properties

DEFINITIONS

Let $f(x)$ be a function of the real variable x, single-valued and possibly complex in value. Then $F(u)$, defined by

$$F(u) = \int_{-\infty}^{\infty} f(x)\exp(-i2\pi ux)dx,\qquad (A.1)$$

is the *Fourier transform* of $f(x)$. It can be shown that a reciprocal relationship then holds,

$$f(x) = \int_{-\infty}^{\infty} F(u)\exp(i2\pi xu)du,\qquad (A.2)$$

so that $f(x)$ is said to be the inverse Fourier transform of $F(u)$. We think of the functions $f(x)$ and $F(u)$ as existing in two different regions or domains, the x-domain and the u-domain, and the Fourier transformation links pairs of functions in these domains. For example, let $f(x)$ be defined by

$$\left.\begin{array}{l} f(x) = 1,\ \ |x| < a/2, \\ f(x) = 0,\ \ |x| > a/2. \end{array}\right\}\qquad (A.3)$$

This is the *rectangle function*, written rect(x/a). Then by elementary integration we find for its transform,

$$F(u) = |a|\operatorname{sinc}(au),\qquad (A.4)$$

where the *sinc function* is defined by

$$\operatorname{sinc}\theta \equiv \frac{\sin\pi\theta}{\pi\theta}.\qquad (A.5)$$

This fundamental pair of functions is illustrated in Fig. A.1, and the relation between them is sometimes written in the form

$$|a|\operatorname{sinc} au \rightleftharpoons \operatorname{rect}(x/a).\qquad (A.6)$$

This notation obscures the distinction between direct and inverse

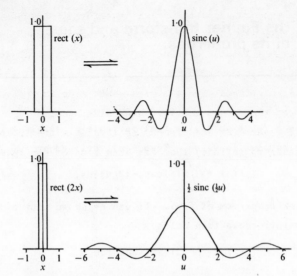

FIG.A.1. The fundamental Fourier-transform pair, the rectangle function, and the sinc function, showing the effect of a change of scale.

transforms but this is often unimportant in physical applications. Fig. A.1 shows, through the scale factor a, how a spreading of one function (a increasing in rect(x/a)) causes the transform to be compressed along the u-axis.

Two-dimensional transforms are defined similarly in terms of functions of two variables,

$$\left.\begin{array}{l} F(u,v) = \iint\limits_{-\infty}^{\infty} f(x,y)\exp\{i2\pi(ux + vy)\}dxdy, \\[2mm] f(x,y) = \iint\limits_{-\infty}^{\infty} F(u,v)\exp\{2\pi i(xu + yv)\}dudv, \end{array}\right\} \quad (A.7)$$

and the theorems given below for one-dimensional functions apply *mutatis mutandis* to two dimensions.

THE DELTA FUNCTION

Fourier transforms can be defined for a great variety of functions, although a discussion of the conditions under which

any given function can have a Fourier transform is beyond
the scope of this Appendix. If $f(x)$ is a constant, say b, the
integral in eqn (A.1) does not converge, and so the transform
of a constant is not defined. However, if in eqn (A.6) the
constant a is very large then the right-hand side is unity
over this large range, and the left-hand side becomes
correspondingly narrower and higher, as in Fig. A.2. In the
limit as a tends to infinity the left-hand side of eqn (A.6)

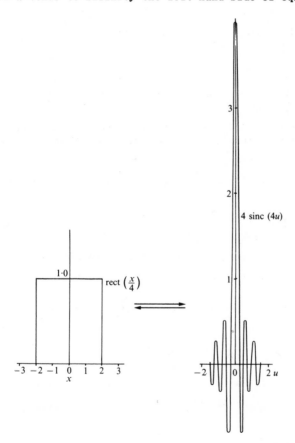

FIG. A.2. Evolution of the delta function as the transform of
a constant. As a increases the transform of rect (x/a) tends to
a delta function. Other pairs of transforms can also be used to
define the delta function by a similar limiting process.

tends to an infinitely narrow and infinitely high spike. This is one way of developing a definition of the *delta function* $\delta(u)$, introduced by P.A.M.Dirac. Thus we have the transform relationship

$$a \rightleftharpoons a\delta(u), \qquad (A.8)$$

where a is a constant. There are many situations in physics where a quantity is constant for a long time or distance, so that its transform is very sharp and narrow, almost a delta function. Thus it is the limiting process implied in eqn (A.8) which is important in physical applications of Fourier-transform theory.

ELEMENTARY PROPERTIES

If we have two pairs of functions which are transforms then any linear combination can make a transform pair,

$$aF(u) + bG(u) \rightleftharpoons af(x) + bg(x). \qquad (A.9)$$

A shift of origin in one domain corresponds to multiplication by a complex exponential in the other domain,

$$\left. \begin{array}{l} f(x + a) \rightleftharpoons \exp(i2\pi au)F(u), \\ F(u - a) \rightleftharpoons \exp(i2\pi ax)f(x). \end{array} \right\} \qquad (A.10)$$

If the above result is applied to the delta function and its transform we obtain

$$\delta(u - u_0) \rightleftharpoons \exp(i2\pi u_0 x), \qquad (A.11)$$

i.e. the transform of a complex exponential is a delta function with shifted origin. This result enables us to write down the transform of a periodic function. Thus if

$$f(x) = a\cos 2\pi u_0 x$$

$$= \tfrac{1}{2}a\exp(i2\pi u_0 x) + \tfrac{1}{2}a\exp(-i2\pi u_0 x)$$

we find from eqn (A.11)

$$F(u) = \tfrac{1}{2}a\delta(u - u_0) + \tfrac{1}{2}a\delta(u + u_0).$$

This can then be extended to a sum of periodic functions, e.g. a Fourier series,

$$f(x) = \Sigma a_n \exp(i2\pi nx),$$

$$F(u) = \Sigma a_n \delta(u - u_n).$$

A useful property of the delta function is that it encloses unit area,

$$\left. \begin{aligned} \int_a^b \delta(u - u_0)\,du &= 1, \quad a < u_0 < b \\ &= 0, u_0 < a \text{ or } b < u_0. \end{aligned} \right\} \quad (A.12)$$

THEOREMS

If $f(x)$ and $F(u)$ are a transform pair then

$$\int_{-\infty}^{\infty} |F(u)|^2 du = \int_{-\infty}^{\infty} |f(x)|^2 dx, \quad (A.13)$$

which is *Parseval's theorem*.

The *convolution* of two functions $f(x)$ and $g(x)$ is defined to be

$$\left. \begin{aligned} f(x) \bigstar g(x) &\equiv \int_{-\infty}^{\infty} f(s - x)g(x)\,dx \\ &= \int_{-\infty}^{\infty} f(x)g(s - x)\,dx. \end{aligned} \right\} \quad (A.14)$$

The *convolution theorem* states that the transform of the convolution of two functions is the product of their transforms,

$$f(x) \bigstar g(x) \rightleftharpoons F(u)G(u),$$

or, more explicitly,

$$\int_{-\infty}^{\infty} \exp(-i2\pi su) \left\{ \int_{-\infty}^{\infty} f(s - x)g(x)\,dx \right\} ds = F(u)G(u). \quad (A.15)$$

Thus convolution in one domain corresponds to multiplication in the other.

The *autocorrelation function* of $f(x)$ is defined as

$$f(x) \; \bigstar \; f(x) \equiv \int_{-\infty}^{\infty} f(x + s)f^*(x)\mathrm{d}x, \qquad \text{(A.16)}$$

where $f^*(x)$ is the complex conjugate of $f(x)$; but sometimes it is more convenient to use a normalized autocorrelation function, in which case the right-hand side of eqn (A.16) is divided by the normalizing constant,

$$\int_{-\infty}^{\infty} |f(x)|^2 \mathrm{d}x.$$

The *autocorrelation theorem* or *Wiener-Khintchine theorem* states that the transform of the autocorrelation of a function is the squared modulus of its transform,

$$\int_{-\infty}^{\infty} \exp(-\mathrm{i}2\pi su) \left\{ \int_{-\infty}^{\infty} f(x + s)f^*(x)\mathrm{d}x \right\} \mathrm{d}s = |F(u)|^2. \quad \text{(A.17)}$$

OPTICAL ANALOGIES

The results given in this Appendix are mathematical expressions of many physical effects. For example, we saw on p.54 that the complex amplitude in the far-field diffraction pattern of an aperture is the Fourier transform of the complex-amplitude distribution in the aperture. In a slightly different context, the same result means that the complex amplitude in the point spread function of a lens is the transform of the complex-amplitude distribution in the exit pupil of the lens (p. 86). Again, the complex amplitude of a plane wave striking an aperture in a plane screen normally is represented by a constant over the aperture, so that the far-field pattern tends to a delta-function shape (i.e. a narrow high peak) as the aperture gets wider. This illustrates the scaling rules given on p.53.

The image of an extended object formed in incoherent illumination is the convolution of the light-intensity distribution in the point spread function of the lens with that in the object. The convolution theorem tells us that in the transform domain this corresponds to multiplying the spatial frequency distribution in the object by the optical transfer function to obtain the spatial frequency distribution in the image (p.89).

The fringe function of a two-beam interferometer is the transform of the intensity distribution in the spectrum of the light, or its *power spectrum*(p(p.97). The autocorrelation function (with time as the variable) of a polychromatic beam of light is, by the autocorrelation theorem, the transform of the square of the power spectrum of the light (p.115).

Note that in the above summary we have omitted scaling and normalizing factors, etc. which would be needed for statements of the results in forms suitable for numerical calculation.

It should be noted also that Fourier transform methods can sometimes lead to dimensionally inconsistent results, particularly in the use of delta functions, since the theory itself is purely mathematical. For instance in Chapter 5 we expressed the image of an extended object as a convolution integral (eqn 5.4) and it can be seen that image and object do not have the same physical dimensions, although both are light intensities. It is possible to reformulate the theory to overcome this, at the expense of some complication, but this is not customary and it is unnecessary in our kind of optics, since we are concerned only with relative light intensities.

References and further reading

M. Born and E. Wolf, (1965), *Principles of optics*. Pergamon Press, Oxford.

D.J.E. Ingram, (1973).*Radiation and quantum physics* (OPS 3). Clarendon Press, Oxford.

H.S. Lipson (ed.) (1972) *Optical transforms*. Academic Press, New York.

F.N.H.Robinson (1973) *Electromagnetism* (OPS 1). Clarendon Press, Oxford.

W.T. Welford, (1962) *Geometrical optics; optical instrumentation*. North-Holland, Amsterdam.

For descriptions of many of the classical interference and diffraction experiments see e.g. R.W. Ditchburn, (1952). *Light*. Blackie, London or R.S. Longhurst (1973). *Geometrical and physical optics* (3rd edn). Longmans, London.

Answers to numerical problems

In some answers only one or two significant figures are given. This indicates that only that precision is physically meaningful, or that the initial data have only that precision.

1.1. 300 m, 300 mm, 300 µm, 300 nm.

1.6. 50 µm, 6×10^{12} Hz.

1.8. Curve 2, 5 per cent; curve 3, 10 per cent.

2.9. $\frac{5}{3}$, -1, $\frac{1}{3}$.

2.10. Focal length $r/2$; image positions -50 mm, 100 mm, 66.7 mm; magnifications 2, -1, $-\frac{1}{3}$.

2.11. (a) $\frac{1}{4\pi}$ W mm^{-2} sr^{-1} (in this solution it is assumed that the total area is 10 mm^2 and the filament radiates uniformly in all directions) (b) 0.03 W.

3.1. 570 m.

3.2. 140 mm^{-1}.

3.6. 0.8 nm.

3.8. 10 m.

3.9. 2 mm.

3.11. 0.5 km.

4.1. (a) 0.985; 0.707; 0.035. (b) 0.970; 0.500; 0.00122.

4.2. (a) 1; (b) 4; (c) 1.987.

4.3. 56.3°, 58.0°, 62.2°.

4.5. 3.4 µm.

5.1. 6.7×10^{-6} rad, 6.7×10^{-7} rad, 1.3×10^{-7} rad (the wavelength is taken as 550 nm).

5.2. 7 µm in diameter; 2×10^{-3} rad mm^{-1}.

5.3. 7 mm.

5.5. 0.5 µm, 500.

5.6. (a) NA 0.25, ×125, (b) NA 1.3. × 1000. (The suggested answers give NA and magnification about 4 times larger than needed for resolution of the dimensions given).

6.1.   (a) 0.546 mm, (b) 0.13 mm; (c) the fringe contrast
       would decrease.

6.2.   0.04 arcsec.

6.5. $\delta\lambda = 0.002$ nm; $\delta\nu = 1.5 \times 10^9$ Hz

6.7.   200 nm to 400 nm.

6.8. 63.5°.

6.9.   5300.

6.10. $F = 30$, $T_{min} = 0.0028$.

6.11. 6×10^5, 0.025 nm.

7.1.   12 mm diameter.

7.2.   6 μm.

7.4.   (a) 7.9; (b) 5.6; (c) 0, but fringes will appear if the
       surface is viewed obliquely.

7.5.   The beams intersect at $\pm 18.45^\circ$ to the normal to the
       grating; 3160 rulings per millimetre.

Index

Abbe, Ernst, 133
aberrations, 35
accommodation, 81
achromatic doublet, 37
adaptation, 81
afocal system, 58
Airy disc, 55
Airy, G.B., 55
Airy pattern, 56, 85
amplitude, 2
angular dispersion, 103
angular frequency, 6
angular magnification, 75
angular resolution limit, 77
angular resolution of the
  eye, 82
anisotropic medium, 68
anti-reflection coating, 108
aperture stop, 33, 81
astigmatism, 35
atmospheric turbulence, 78
autocorrelation function, 142
autocorrelation theorem, 142

beam-expander, 57
beam-splitter, 40, 95
birefringence, 68
birefringent crystal, 70
Brewster angle, 67

calcite, 68
camera objective, 98
Cassegrain telescope, 79, 80
central maximum, 57
chromatic aberration, 36, 80
circular aperture, 55
circular fringes, 95
circular polarization, 65, 69
coherence, 44, 91
coherence length, 15, 97
coherence patch, 92
coherence time, 15, 97
collimator, 29
complex amplitude, 4, 6, 7

complex exponential
  notation, 4
compound microscope, 84
concave grating, 102
conjugate distance
  equation, 28
conservative field, 25
contrast, 45, 92
contrast transfer, 89
convergence angle, 32
convolution, 88, 141
convolution theorem, 141
cornea, 81
corrector plate, 79
correlation spectroscopy, 116
cosine Fourier transform, 97
critical angle, 26
crown glass, 37
curvature, 27

de Broglie wavelength, 84
delta function, 138
detector, 80
detectors of light, 4
differentiating filter, 135
diffracted beam, 100
diffracted beam, first
  order, 122
diffracted beam, of Mth-
  order, 101
diffracted beam, zero
  order, 125
diffracting screen, 57
diffraction, Chapter 3, 45
diffraction at an aperture, 48
diffraction at an edge, 46
diffraction grating, 99
diffraction gratings,
  holographic, 129
diffraction-limited pin-
  hole, 92
diffractometer, 57
dispersing prism, 98
dispersion, 36, 103

dispersion curve, 36
dispersion of a prism, 103
double refraction, 68

effective number of interfering
  beams, 113
electric field, 1
electromagnetic spectrum, 1
electron lens, 84
electron microscopy, 84
elliptically polarized light,
  65
energy, 3
entrance pupil, 76
entrance slit, 103
equation of wave motion, 46
exit pupil, 76
exit slit, 103
extended object, 27, 85, 87
extended source, 43
eye, 80
eyepiece, 74

Fabry-Perot interferometer,
  108
Fabry-Perot transmission, 111
Faraday effect, 71
far field, 29, 52
far-field diffraction, 50, 52
  55
Fermat's principle, 24
finesse, 112
flint glass, 37
flux density detectors, 11
focal length, 28
focal plane, 74
focal ratio, 80
Fourier plane, 132
Fourier transform, 54, 89, 137
  Fourier-transform spectroscopy, 97
Fraunhofer diffraction, 52
free spectral range, 113
frequency, 1
Fresnel, A., 48
Fresnel diffraction, 53
fringe function, 96
fringe maximum, 59

Gabor, D., 124
Gaussian approximation, 30
Gaussian optics, 18, 34
geometrical optics, Chapter
  2, 18
geometrical wavefront, 21
grating monochromator, 102
grating spectrograph, 102

hologram interferometry, 127
holography, 121
holographic reconstruction,
  121
Huygens, 48
Huygens' secondary wavelets,
  48

Iceland spar, 68
images, 26
impulse response, 88
infinity, object at, 29
infrared, 8
in-plane displacement, 129
intensity, 8
interference, Chapter 3
interference field, 5
interference fringe, 39
interference pattern, 16
interference in polychromatic
  light, 43
intermediate image, 32
intersecting plane waves, 39
invariant, 33
iris, 81
iris diaphragm, 34

Kerr effect, 71
Kirchhoff, G, 48

Lagrange invariant, 32
laser, 119
laser light, 15
latent image, 10
least action, principle of
  25
light, descriptions of, 17
light-gathering power, 74,
  103
light gathering power for
  visual purposes, 76

linear response, 11
logarithmic response, 82
lumen, 13
luminance, 34

magnesium fluoride, 108
magnification, 30, 74
magnetic field, 3
Maupertuis, P.de, 25
mechanical stress 70
mercury lamp, 95
Michelson's interferometer, 94
microscope, 82
microscope objective, 84
minimum deviation, 104
minimum resolvable separation,
   134
mirror objectives, 80
modulation, 89
momentum, 25
monochromatic beam, 14
monochromator, 102
multi-element lenses, 31
multiple-beam interference,
   107, 108
multiplexing, 114

near-field, 53
Newton's rings, 43
Newtonian telescope, 79, 80
Nicol prism, 63
nonlinear optics, 4
nonlinear response of eye, 41
numerical aperture, 85

objective, 53
objective lens, 74
oil film, 42
oil-immersion, 85
optical glass, 36, 80
optical images, 26
optical path length, 24, 26
optical transfer function, 89
optic axis, 69
order of interference, 42, 112
orthogonally polarized beams,
   72
overlapping orders, 113

paraboloidal mirror, 79
paraxial optics, 18
Parseval's theorem, 141
partial coherence, 93
pencil of rays, 21
persistence of vision, 11
phase shift, 5
photocathode, 10
photocell, 11
photoelectron, 10
photographic emulsion, 10
photometry of optical systems,
   76
photomultiplier, 11
photon, 9
photo-resist, 129
plane grating, 105
plane of polarization, 62
point image, 39
point spread function, 75, 77,
   83, 86
polarization, Chapter 4
polarization and interference
   71
polarized light, 63
polarizer, 63
'Polaroid', 70
polychromatic field, 13
power, 3
power density, 3
principal focus, 29
principal planes, 31
prism spectrograph, 98
probability density, 59
pupil function, 86

quantum detection process, 10
quantum efficiency, 12
quarter-wave plate, 69

radio waves, 3
randomness in a light beam, 14
ray, 21
rectangle function, 137
red cadmium line, 97
reference point, 123
reflecting collimator, 102
reflection factor, 66
reflection grating, 101

reflection, law of, 22
refraction, law of, 22
refractive index, 23
resolution, 76
resolving power, 74, 105
resolving power of a prism, 106
resolving power, spectroscopic, 105
response time, 16
retardation, 70
retina, 81

saccharimetry, 71
scalar wave, 18
scalar wave theory, 47
scanning interferometer, 112
Schmidt camera, 79
Schmidt telescope, 80
secondary sources, 91
secondary wavelets, 48
simple harmonic wave, 15
sinc function, 52, 137
sinusoidal grating, 89
Snell's law, 23, 25
spatial filtering, 130
spatial frequency, 88
speckle, 119
spectral-sensitivity curve, 12
spectrograph, 99
spectroscopic grating, 100
spectrum line, 104
spurious real image, 126
stabilized laser, 15
state of polarization, 62
stationarity, 24
stellar interferometer, 93
superposition, 4

Taylor, G.I., 58
telescope, 74
temporal coherence, 94
thermal detection process, 10
thermal energy, 10
thermal source, 9, 91
thin lens, 26, 27
time-averaged power, 4
time-response, 11
total flux detectors, 11
total internal reflection, 26

transmission grating, 121
two-beam interference, 7, 45
two-dimensional transform, 138

ultraviolet, 9
unpolarized light, 65

virtual-image reconstruction, 126
visibility, 45
visual observation, 76

wavefront, 4, 6, 21
wavelength, 1
wave-number, 6
wave-train, 15
wave-vector, 6
white light, 14
Wiener-Khintchine theorem, 142

Young, Thomas, 91

zero order, 100
zero path difference, 97